Off-Grid Survival Projects

An Expert DIY Guide to No Grid Self-Sufficiency, Food Security, and Renewable Energy with Real-World Solutions

HARPER MOSS

Feel free to join our mailing list for updates and information about other upcoming books by Harper Moss.

HARPER MOSS

Copyright © 2025 by **Harper Moss**

---------------------------------- **ALL RIGHTS RESERVED** ----------------------------------

No portion of this book may be reproduced in any form without written permission from the publisher or author except as permitted by U.S. copyright law.

Table of Contents

INTRODUCTION 1

SECTION 1: WATER SECURITY & SUSTAINABILITY 6
    Chapter 1: Understanding Off-Grid Water Needs 7
    Chapter 2: Water Collection Systems 11
    Chapter 3: Water Filtration & Storage 21

SECTION 2: OFF-GRID ENERGY INDEPENDENCE 29
    Chapter 4: Assessing Energy Needs & Reducing Consumption 31
    Chapter 5: Solar Energy for Off-Grid Power 35
    Chapter 6: Wind & Hybrid Energy Solutions 43

SECTION 3: FOOD SECURITY & SUSTAINABLE LIVING 50
    Chapter 7: Off-Grid Cooking Essentials & Fuel Sources 53
    Chapter 8: Food Gathering, Hunting & Preservation 59
    Chapter 9: Livestock Raising & Beekeeping 65
    Chapter 10: Advanced Gardening & Permaculture 73

SECTION 4: HEALTH, HYGIENE & EMERGENCY PREPAREDNESS 81
    Chapter 11: Off-Grid Hygiene & Waste Management 83
    Chapter 12: Natural Remedies & First Aid 89

SECTION 5: SECURITY & LONG-TERM RESILIENCE 94
    Chapter 13: Off-Grid Security & Home Defense 97
    Chapter 14: Mental Resilience & Community Building 103

FINAL SECTION: OFF-GRID READINESS & DIY PROJECTS 109
    Chapter 15: Final Off-Grid Readiness Checklist & Action Plan 111
    More Real-World Examples & Case Studies 121
    The Last Word about Off-Grid Living 131

DIY Projects

DIY PROJECT 1:
Build a Simple Water Usage Calculator 10

DIY PROJECT 2:
Build a Rainwater Collection System 18

DIY PROJECT 3:
Construct a Sand and Charcoal Water Filter 26

DIY PROJECT 4:
Build an Energy Usage Tracker 33

DIY PROJECT 5:
Install a Small Solar Panel System 40

DIY PROJECT 6:
Build a Simple Wind Turbine 47

DIY PROJECT 7:
Build a Rocket Stove 56

DIY PROJECT 8:
Build a Survival Snare Trap 63

DIY PROJECT 9:
Build a Simple Beehive (Langstroth Design) 71

DIY PROJECT 10:
Build a Raised Bed Garden 78

DIY PROJECT 11:
Construct a Composting Toilet 87

DIY PROJECT 12:
Make a Natural Herbal First Aid Kit 92

DIY PROJECT 13:
Install a Simple Motion Sensor Alarm 101

DIY PROJECT 14:
Make an Emergency Barter Kit 106

DIY Projects *OTHER*

DIY Rainwater Harvesting (Step-by-Step)	12
DIY Sand and Charcoal Filters	22
DIY Solar Power Setup	36
Setting Up a Small Wind Turbine (DIY)	44
DIY Smokers & Dehydrators for Food Preservation	54
DIY Smokers — Preserving with Smoke and Heat	54
DIY Dehydrators — Harnessing Heat and Airflow to Preserve Your Harvest	55
DIY Beehive Construction	69
Raised Bed Gardening (DIY)	74
DIY Perimeter Fencing & Concealment Landscaping	99
Building DIY Perimeter Fencing	99

INTRODUCTION

Understanding Off-Grid Living: Myths vs. Reality

Stories of living off the grid are abundant these days. Everyone seems to have a distinct definition of living off the grid, based on everything from web videos to television series. Some picture themselves completely alone in a log cabin tucked away in the woods. Others see unending bright days with solar-powered power, eating garden-fresh food, and living independently.

People rarely think that off-grid living is what it is. It is neither a romantic getaway from contemporary life nor a never-ending survival struggle. Instead, it is a way of life centered on education, diligence, problem-solving, and fortitude.

Simply said, living off the grid means not depending on public amenities such as sewage, electricity, or water. However, living off the grid does not entail isolating oneself or becoming a recluse. Many people who live off the grid use technology to stay in touch, create vibrant communities, and even operate businesses.

There are many myths to dispel. One of the most significant is that living off the grid can be costly or extremely inexpensive. It depends on your decisions, territory, and how you construct your systems. You can save money on specific tasks. Others will need an initial financial outlay. Knowing the difference between what you need and what you want is crucial.

The idea that living off the grid is easy is another misconception. The reality is that managing your own food, water, and electricity systems is not easy, at least not initially. However, it has become a way of life through practice and knowledge. You begin to view the difficulties as chances for personal development rather than barriers.

This book aims to help you overcome misconceptions and position yourself for success as you navigate the realities of off-grid living.

Whom This Book Is For (Beginners to Advanced Preppers)

Anyone who aspires to live with greater control over their finances should read this book. You will find something helpful here whether you are already heavily involved in preparation or inquisitive.

This book serves as a beginning point for those new to homesteading or who have always aspired to become more independent. You don't require any expertise or abilities. Each project is intended to provide you with a strong foundation and guide you through the fundamentals.

You are most likely prepared to take on more challenging tasks after completing a few assignments. This book will teach you new techniques that will push you beyond the fundamentals and help you avoid common mistakes.

There are tasks and ideas for seasoned homesteaders and preppers that will push you to improve your setup. Perhaps you already have a garden or solar power. With strategies that increase your productivity, fortify your defenses, and broaden your understanding, this book goes beyond.

The objective is straightforward, regardless of where you begin: develop the abilities and frameworks that provide you autonomy and self-assurance.

Overview of Essential Off-Grid Projects

Off-grid living covers a lot of ground, but at the core, it is about meeting your basic needs without depending on public utilities or outside systems. The good news is once you understand the main categories of off-grid life, everything else starts to make sense.

In this book, you will find practical projects designed to meet the following needs:

- **Power Generation** — Learn how to build your own solar systems, explore wind power, and set up battery banks that keep your lights on when others go dark.

- **Water Collection and Purification** — Harvest rainwater, dig a well, filter water naturally, and create systems that keep clean water flowing year-round.

- **Food Production and Preservation** — Grow your own vegetables, raise small livestock, forage wild edibles, and learn to preserve your harvest through canning, drying, and more.

- **Shelter and Heating** — Build simple, strong structures, improve insulation, and create heating systems that keep you warm through the coldest winters.

- **Sanitation and Waste Management** — Design composting toilets, handle greywater safely, and even turn waste into usable energy.

- **Communication and Security** — Protect your home, set up emergency communication systems, and create a safe environment for your family.

Each project is designed to teach you how to survive and live well off the grid.

How to Use This Book (DIY Project-Based Approach)

The purpose of this book is to be helpful. This book does not include abstract concepts, nor is it a survival guide. Instead, it is a series of do-it-yourself projects that guide you through creating actual systems.

There are tasks in each chapter that you can do with simple tools and a little time. Each chapter focuses on a different aspect of off-grid living, such as food, power, or water. Every project consists of:

- A clear explanation of why it matters
- A list of materials and tools
- Step-by-step instructions anyone can follow
- Estimated time and cost
- Tips to avoid common mistakes

The projects don't have to be completed in that order. Choose the one that will help you overcome your most significant obstacle now. Perhaps you need a better water system, but they already have solar power. Maybe you would like to begin with a modest food preservation project. In any case, you can follow your own path with this book.

Use it as a field guide that you consult when building time comes.

Common Pitfalls in Going Off Grid and How to Avoid Them

Many individuals dive into off-grid living without fully comprehending the implications. That's how dreams become headaches and initiatives are shelved. However, you can steer clear of most of those blunders by taking advice from individuals who have been there before.

The following are a few of the most typical pitfalls:

Thinking it will be inexpensive

Living off the grid can save you money, but not immediately. Water collection and solar systems are expensive upfront. Avoid wasting your money on unnecessary purchases. This book aids in making informed decisions.

Using a single system

Solar power is fantastic for up to a week of cloud cover. Backups include wood burners, generators, and rainwater collection in a clever configuration. Layers of systems are necessary to keep you informed.

Ignoring the learning curve

Purchasing equipment is simple. The true work comes in knowing how to use it and fix it. Building skills, not merely purchasing equipment, is the focus of this book.

Disregarding local regulations

You must be aware of the laws even if you are the landowner. Zoning, building rules, and water rights are all important. Verify before beginning any construction or excavation.

Making the way of life more romantic

Living off the grid is more than simply fresh eggs and sunsets. There are difficult days, inclement weather, and malfunctions. If you're ready, you'll deal with it when it happens.

Preparing Mentally, Financially, and Logistically for the Transition

Preparing yourself mentally, financially, and strategically is crucial before installing solar panels or drilling your first well.

Mental readiness

You will be put to the test in this life. Some days will be indescribable in their rewards. On other days, it will be tiresome and frustrating. Understanding that both are a part of the process is crucial. Honor the little victories. Make fun of the failures. Continue.

Getting ready financially

You don't have to be wealthy to live off the grid, but you must know where your money goes. Specific projects are well worth the investment. Others can wait. This book tells you where to save money and helps you prepare for the significant costs.

Preparation of logistics

Off-grid living is like managing a small farm, power plant, and workshop simultaneously. You require equipment, materials, contingency plans, and expertise. Additionally, you must be familiar with your land, including its sunniest locations, water flow patterns, and soil composition. Everything else is made easier by careful planning.

Case Study: A Real-Life Off-Grid Transition Story

I would like to present John and Sarah Miller to you. The noise, the expenses, and the sense of being stuck were all things they had grown weary of in the city. Thus, they purchased twenty acres of wooded land devoid of buildings, water, and electricity. Only dreams and trees.

They almost went bankrupt in the first year.

That summer, they well dried up. Their initial solar setup was insufficient to meet even the most basic requirements. Their garden didn't work out. What about winter? They almost returned to the city during the winter.

They didn't give up, though.

They gained knowledge. They constructed a tiny wind turbine as a backup and added rainwater collection. They discovered ways to extend their electricity and improve the insulation in their cabin. They built a greenhouse that could withstand the snow and began rearing hens.

The Millers are still alive and well today. Most of their food is grown there. They produce their own energy. They lead independent lives. Their success resulted from patience, a readiness to learn, and a determination to keep going when things became difficult, not chance or wealth.

There are several stories like theirs. This life can be yours if you take it seriously.

Living off the grid is not a pipe dream. It's real. It's challenging. It's lovely. And all your hard work is worthwhile.

This book is your manual whether you wish to live entirely off the grid or simply increase your level of independence. You get closer to living a life where you depend on yourself rather than on unreliable systems with each project you take on and the skills you acquire.

One project at a time, please. Start small if necessary. Simply begin.

The voyage has begun. Let's go to work.

Section I:
WATER SECURITY & SUSTAINABILITY

Chapter 1: Understanding Off-Grid Water Needs

How Much Water Do You Need?

Knowing how much water you use, rather than just where to get it, is one of the first things you need to know while living off the grid. Water idly flows through a suburban home or city. Off the grid, it becomes a resource that you value, protect, and monitor regularly.

For drinking purposes alone, each person needs at least two to three liters of water every day. But that's only the beginning. Cooking, cleaning, bathing, caring for animals, and growing food all add up. Even with rigorous conservation, a small family can easily use fifty to one hundred gallons of water daily.

Here is what typically pulls the most water:

- Drinking and cooking
- Bathing and laundry
- Washing dishes and cleaning
- Watering gardens and livestock

Some days, it is the garden that drains your supply. Other times, a string of hot days pushes everyone to drink more. Seasons matter, too. Winter brings less water needs for plants, but you might end up melting snow. Summer pulls water faster than you think.

Understanding your average daily use is the foundation of building a sustainable water system. Most people are shocked once they start tracking it. The best way to avoid overbuilding — or worse, running short — is to measure your real needs before you build a single system.

Emergency Water Sources in a Crisis

No matter how good your setup is, things can go wrong. Pumps fail. Tanks crack. Dry spells drag on. The real test of off-grid living is what you do when your main water supply suddenly disappears.

Nature offers a few backup options — if you know where to look:

- In an emergency, water can be obtained from lakes, rivers, and streams. However, they also provide hazards like chemical runoff, bacteria, or parasites. It is necessary to properly filter and purify any surface water.
- Rainwater collection is one of the smartest systems to have. A well-designed catchment setup keeps you afloat during dry times, but only if you maintain it.
- Snow and ice are the last resorts. They will give you water if you melt enough of them, but it is hard work for little yield.
- Collecting Dew is a survival trick — gathering it off tarps or plants early in the morning is more of a supplement than a true supply.

The secret is to know your land. Walk it. Learn where the water gathers, what dries up first, and where you might pull from in an emergency. Do not wait until you are thirsty to figure this out. Plan now while you have the time.

Modern Off-Grid Water Tech: UV Pens, Ceramic Filters, Desalination

Technology has changed the off-grid water game. What used to be labor-intensive work can now be handled with simple tools — if you know what to choose.

- **UV Water Purifiers:** Small, portable pens use ultraviolet light to kill bacteria, viruses, and parasites. Perfect for emergency use or small batches, they are fast and effective. But they rely on batteries or solar power.
- **Ceramic Filters:** These simple filters have stood the test of time. They remove sediment and pathogens, often lasting for thousands of gallons. They are easy to clean, affordable, and great for daily use.
- **Desalination Units:** Turning salt water into drinking water is no small feat, but portable desalination tech has come a long way. Still, these units remain expensive and energy-hungry, making them best for specific locations near the ocean.

Using a single tool is seldom the best course of action. Although it can fail, technology can be helpful. The secret to keeping your family safe in an emergency is to build layers or several ways to store and filter water.

DIY Project 1:
Build a Simple Water Usage Calculator

You must know how much water you use before drilling, excavating, or erecting water tanks. Most individuals make assumptions. It's a mistake. This straightforward project lets you organize your system based on facts rather than hope by providing concrete data with which to work.

What you need:

- A notebook
- Measuring containers

How to start:

You don't have to be flawless; just be honest. Keep a record of every drop your household uses for three whole days. Write down what it was used for and how much water you used.

Create categories as you track:

- Drinking
- Cooking
- Washing dishes
- Laundry
- Bathing
- Watering plants or animals

After three days, you will notice trends. Perhaps the garden is absorbing more than it should. Perhaps laundry is taking up a lot of space. Determine your average daily consumption after you have your totals.

You can calculate your weekly requirement by multiplying that figure by seven. If you'd like, plot it by season or month. This easy exercise transforms your perspective on water and equips you with the information to properly size your collection, filter, and storage tanks.

This is the first step in intelligent off-grid living: figuring out what you need and working from there.

Chapter 2: Water Collection Systems

You cannot ignore water collection when you live off the grid. Since everything stops without water, it becomes one of the most crucial systems you will ever construct. Your garden becomes parched. Your animals become thirsty. Even easy chores like cleaning and cooking become difficult.

The good news is that if you know where to look and how to catch it, you can find plenty of water in nature when It rains. Under your feet, groundwater runs. In the winter, snow accumulates. If you are shrewd enough to extract it, moisture can even be found in the air.

The secret is to learn how to make the most of your climate and terrain. A family living on arid, broad plains could not understand what works for someone deep in the forest. Each property is unique. Every off-grid residence requires a solution tailored to its own set of difficulties.

This chapter will examine the most popular off-grid water collection methods. We'll go over the benefits and drawbacks of every option, from straightforward rainfall collection devices to wells that draw from subterranean supplies. You will also learn why snow and ice should be included in your water plan and how greywater, typically thrown away, can become an important system component.

Relying solely on one source is not the aim. The goal is to have many security levels and a backup plan in case one system fails. That is what sustains you during freezing, droughts, and all other conditions.

DIY Rainwater Harvesting (Step-by-Step)

Installing a rainwater collecting system is one of the most intelligent off-grid projects you can do. It is easy to use, efficient, and, once constructed, continues to function each time the sky opens. This method transforms your roof into a reliable water source for your home, garden, and cattle.

You don't have to go fancy or spend thousands. With some preparation and simple supplies, you can create a system that will work effectively for you for many years.

Step 1 — Choose the Right Collection Surface

The simplest place to start is on your roof. The first thing you need is a surface to catch the rain. Metal roofs are perfect because they shed water rapidly and are easy to maintain. Shingle roofs can also be effective, but if you intend to drink the water, you will need a better filtration system because they can occasionally introduce pollutants and grit.

Find the areas with the most rainfall by strolling around your house or shed. The greatest place to start is typically on larger, steeper roofs because they collect more water.

Step 2 — Install Gutters and Downspouts

If you don't already have gutters, it's time to install them after selecting your roof. Rainwater is directed to your storage tanks via gutters from the roof. To ensure easy water flow, ensure the gutters are clean, securely sealed, and angled slightly toward your downspouts.

Where the water will exit the roof, connect the downspouts. Water will be directed into your collection system by these.

- Use screens or leaf guards to stop debris from clogging the gutters.
- Make sure downspouts are large enough to handle heavy rainfall.

Step 3 — Add a First-Flush Diverter (Optional but Smart)

Dust, leaves, and bird droppings are washed down when the rain first reaches your roof. You may catch that initial, unclean flow and move it away from your storage tank with a first-flush diverter.

It is a straightforward add-on, often consisting of a vertical pipe that fills up before cleaner water enters your storage. After the tank is filled, the remaining rainwater enters the tank without going through the diverter.

This small action significantly impacts, particularly if you intend to use the rainwater for drinking or cooking.

Step 4 — Set Up Your Storage Tanks or Barrels

It's time to get that water now. True magic happens in your storage container, transforming free rainwater into something you can use whenever possible.

Select barrels or tanks made for storing water that are suitable for food. Your space and the amount of rain you receive will determine the size. Some people begin with a couple of barrels that hold fifty gallons. Some people use a thousand-gallon tank.

- Set the tank on a level, sturdy base; compacted gravel, concrete slabs, or cinder blocks are good options.
- If possible, keep your storage area shaded. The growth of algae within the tank is accelerated by sunlight.

When your tank fills up, install an overflow outlet close to the top so that extra water can securely drain away.

Step 5 — Add Basic Filtration

Before using water, especially for drinking or cooking, it is wise to pass it through a simple filter, even if it appears clean. Simple screens that capture leaves and insects may be sufficient for use in gardens or with cattle.

For household use, consider adding:

- A fine mesh screen at the tank inlet
- A sediment filter to catch fine particles
- A charcoal filter for better taste and odor control

Remember, you can continually improve filtration later. Start simple, test your water, and build from there.

Step 6 — Install a Spigot or Pump for Easy Access

Make your water easy to use after you've stored it. A spigot at the bottom of your tank allows you to attach a hose or fill buckets.

A tiny pump helps transport water where you need it, especially if you intend to run it into your house for larger systems or tanks that are low to the earth.

Step 7 — Test the System and Maintain Regularly

After everything is set up, watch your system function and wait for a decent rain. Look for weak points where water can pour out, such as leaks or obstructions.

Rainwater systems require routine maintenance:

- Clean gutters and screens often
- Check the first-flush diverter after heavy rains
- Inspect the tank for cracks or leaks
- Flush out sediment build-up every few months

How to Dig and Maintain a Well

Having your own well signifies ultimate independence for many off-grid residences. No one can turn off or meter the potent force of digging into the earth and extracting pure water. However, well excavation is a significant undertaking that requires preparation and perseverance and, if done well, will pay off for years to come.

Knowing the procedure helps you prevent expensive errors and maintain your water flow over time, whether you work with experts or try a manual method.

Planning Your Well — Where to Dig and What to Expect

Spend some time planning before you start. A well is your lifeblood, not just a hole in the earth. Pick the spot wisely.

- **Distance from Contaminants:** Your well should be far from septic systems, livestock pens, or anything that might leak waste into the ground. Forty to fifty feet away is often recommended, but local regulations might require more.
- **Natural Drainage:** Pick a spot where water runs away, not toward the well. Avoid low spots where surface water collects — this reduces the risk of contamination.
- **Access for Equipment:** If you hire a drilling crew, ensure they can reach the site with their rigs.

Before digging, check your local laws and get permits if needed. Water rights vary by location, and some areas regulate well depth, construction, and groundwater use.

Types of Wells — Which One Fits Your Land and Budget

There are a few ways to dig well, depending on your budget, location, and how deep the water is:

1. Dug Wells:

Hand-dug wells are old-school — shallow, wide holes lined with stone, brick, or concrete. They work where the water table is high but can run dry during droughts. These days, they are rare but still valuable in certain areas.

2. Driven Wells:

Sand-point wells are also created by driving a small-diameter pipe with a screened end straight down into sandy or loose soil. This method is affordable and great for temporary or backup water, but it only works where the water table is shallow.

3. Drilled Wells:

The most reliable option, especially for deep water sources. A professional drilling rig bores hundreds of feet if needed. These wells tap into aquifers, providing steady water year-round. They cost more upfront but deliver the best long-term security.

If your land is rocky or your water table is deep, drilling is often your only option.

The Drilling Process — What Really Happens

If you choose to drill, here's what you're going to expect:

- A drilling crew arrives with heavy equipment and bores down until they hit water-bearing rock or soil.
- The hole is lined with steel or plastic casing to prevent collapse and protect the water from contamination.
- A screen is installed at the bottom to filter out sediment.
- Finally, the well is capped, sealed, and tested.

Although depth solely depends on your soil and water table, most contemporary wells are between 100 and 500 feet deep.

After drilling, a pump—above-ground or submersible—is required to raise the water level. Although manual hand pumps can be built as a backup, solar-powered pumps are excellent for off-grid installations.

Testing and Tasting — Know What's in Your Water

Just because water is clear does not mean it is clean. Test your well water before drinking — and keep testing regularly.

You are looking for things like:

- Bacteria
- Nitrates
- Heavy metals
- Mineral content like iron or sulfur

Water tests can be done through local labs or with home kits, but lab testing is more thorough. If you find problems, filtration systems can usually fix them, but knowing the front is better than risking your health.

Maintaining Your Well Keep It Flowing and Safe

Wells are challenging, but they are not set-it-and-forget-it systems. Regular maintenance keeps them running strong for decades.

- **Inspect the wellhead:** Make sure the cap is sealed tight. A cracked or loose cap is an open door for insects, animals, and dirty water.
- **Keep the area clean:** No chemicals, fuel, or waste near your wellhead.
- **Flush the system:** Over time, sediment builds up. Flushing keeps your pump from overworking and your water clean.
- **Service the pump:** Pumps work hard. Regular check-ups prevent sudden failures — especially in winter when repairs are more challenging.

Consider testing your water once a year or anytime the water smells, tastes strange, or becomes cloudy.

Greywater Recycling for Sustainable Use

Living off the grid means that every drop of water counts. You put much effort into gathering, storing, and maintaining it. Therefore, wasting water that could still be useful is the last thing you want to do. Greywater recycling, a straightforward but frequently disregarded technique, can help you extend your water supply without compromising comfort or hygienic conditions.

The water that runs off your sinks, showers, baths, and laundry is known as greywater. It's not sewage, but it's also not fresh. There is no human waste, only soap, dirt, or grease residue. Greywater in most homes goes directly into the sewer or septic system, where it is never seen again. However, it is far too precious to spend outside the grid.

Rethinking how water moves through your house is the goal of greywater recycling. You catch it, clean it if necessary, and offer it a second chance at life rather than utilizing it once and then discarding it. For things like watering your garden or trees and bushes, greywater is ideal. Some systems use it to rehydrate compost piles or flush toilets.

The first step in setting up greywater recycling is to separate it from blackwater, the water from toilets, and other sources that transport human waste. This cannot be negotiated. Combining the two makes things more complicated and poses a health risk. After being separated, greywater is gathered via a separate piping system and sent either straight outside or to storage tanks for future use.

For most greywater applications, a complex filtration system is not necessary. Simple filters that capture hair, lint, and big debris are sufficient, mainly if irrigation is your primary objective. Before releasing greywater into the earth, some off-grid families take it further and construct tiny artificial wetlands or sand filters where the water flows through plants and gravel. In addition to providing habitat and aesthetic appeal for the home, the plants and natural materials further purify the water.

After reusing greywater, it's critical to consider what ends up in your drains. Traditional cleaning products, detergents, and soaps are frequently loaded with bad chemicals for the environment, plants, and soil. Switching to plant-safe, biodegradable materials is an easy way to improve the safety and efficiency of your system.

Greywater recycling can be as easy or complicated as you like. Some people connect their laundry to the garden directly with a simple conduit. Others set up holding tanks, pumps, and filters to control bigger systems. In either case, the objective remains: minimize waste, safeguard your freshwater resources, and maximize each gallon.

The simple but necessary is maintenance. If you are left alone, soap scum, lint, or grease can clog greywater lines. A brief inspection once a month or so keeps everything in order. Applying the water to the soil rather than the plants when using it on food crops is preferable to prevent possible contamination. Ornamental plants, berry bushes, and fruit trees adore and flourish in greywater.

Greywater systems can distinguish between a garden that struggles and one that thrives in arid areas. Systems can drain and winterize in colder climates, ready to be restarted when the ground thaws.

Greywater recycling is sustainable and wise. It makes a resource out of what most people consider rubbish. It lessens water trucking, supports the surrounding land, and eases your rainwater or well system burden.

It is challenging to fathom returning to this way of thinking about water after getting used to it. As it should, water becomes a part of a cycle, moving through your house, garden, and back into the ground.

Rainwater Collection: Legal & Environmental Considerations

Rainwater collection seems easy and natural, and it is. However, you must know local laws, regulations, and environmental issues before constructing your system. If not well planned, a project that begins with good intentions may occasionally become mired in red tape or do unintentional harm. Unbelievably, rainfall is regarded as a component of the local watershed in various regions, including some U.S. states and counties. It is officially regulated since every drop that falls eventually reaches rivers, lakes, or subterranean aquifers, supporting communities, crops, and ecosystems. Limiting or managing water removal from the natural cycle, including from your roof, is possible. Large collection systems may need permits in some areas.

Others restrict its storage capacity or its intended usage. Some locations promote rainwater collection and even provide tax credits or subsidies for installing authorized systems. The important thing is to verify long before you install gutters or purchase tanks. Contacting your area's county clerk, environmental agency, or extension office is worthwhile. They can assist you to steer clear of expensive blunders and direct you to the appropriate rules. Certain places have different regulations depending on whether you use the water for domestic purposes, gardening, or pets. Rainwater collection has environmental and legal obligations. By taking water away from the natural processes that replenish aquifers, feed streams, and maintain ecosystem health, you do more than just take it for yourself. Gathering rainwater is still a good idea. It just indicates that you should carefully plan your system. Nearby areas can become arid due to large systems that collect significantly more water than is necessary. Overflow drains with a poor design could flood nearby properties or create erosion. Over time, even the materials you employ, such as specific coatings or plastics, might release pollutants into the nearby soil or water. It's a balanced, smart rainwater system. It meets your requirements without taking things from the environment. It uses safe, long-lasting materials, manages overflowing properly, and progressively returns unused water to the ground. Rainwater collection is generally permitted and encouraged. However, you may create a system that benefits you and the environment by learning your property's regulations and natural rhythms. Rainwater collection, when done correctly, is a sustainable way to maintain your off-grid lifestyle while honoring the water cycle on which everything depends.

Water Collection Systems

DIY Project 2:
Build a Rainwater Collection System

Building your own rainwater collection system is one of the best off-grid tasks. When it rains, you'll know that the water is helping you by filling your barrels and being ready for the garden, your pets, or even your home once it has been properly filtered.

With the correct equipment and a little preparation, this system is easy to assemble over the weekend.

What You'll Need

Large storage containers or food-grade barrels are required for a basic arrangement. These lessen the possibility of chemicals leaking into your supply and are safe for keeping water. Mesh screens keep trash and leaves out of your barrels, and PVC pipes help direct water from your gutters to them.

Additional useful materials include connections, a spigot, sealant, and a strong base to lift the barrels off the ground.

Step 1 — Choose Your Collection Point

Pick the section of your roof that gets the most runoff. A metal roof is ideal because it sheds water quickly and cleanly, but any solid roof will work. Install or clean your gutters so water can flow freely toward the downspout.

Plan where your barrels will sit. Raising them on cinder blocks or a wooden platform makes it easier to use gravity when filling buckets or running a hose later.

Step 2 — Prep Your Barrels

Drill a hole near the bottom of each barrel where you'll install the spigot. This gives you easy access to the water without dipping a bucket in from the top.

Seal the spigot tightly to prevent leaks. Some people also drill an overflow hole near the top of the barrel to allow excess water to drain away safely once the barrel fills.

Step 3 — Connect Your Downspout

Cut your downspout so it lines up neatly with the top of your barrel. Use PVC pipes or flexible tubes to guide the water from the gutter directly into the barrel opening.

At this point, it helps to add a fine mesh screen over the barrel opening. This keeps out leaves, twigs, and bugs while allowing water flow.

For even cleaner water, you can add a simple first-flush diverter — a vertical pipe that fills first and catches the dirtiest water from the initial runoff.

Step 4 — Secure Everything and Test

Once everything is connected, double-check that your pipes and seals are snug. If you can, wait for light rain to test the system. Watch the water flow and ensure it fills the barrel, not spilling over where it should not.

Adjust your system as needed. Adding another barrel in a series is a smart way to increase capacity and avoid overflow if the flow is too strong.

Step 5 — Use and Maintain

Your system is prepared to begin working for you now that it is in place. Water from your rain barrels can be used for gardens, livestock, cleaning tasks, or filtered further for household use.

Just keep in mind that upkeep is important. Regularly check the screens and clean your gutters. Watch the barrels for leaks or the growth of algae. If sediment accumulates over time, flush the system.

Chapter 3: Water Filtration & Storage

Gathering water is just half the fight. Once that water is in your barrels, tanks, or buckets, the true problem starts because not all the water that flows from the ground or falls from the sky is fit for human use. One of the most important resources in any off-grid system is clean, safe water, and how you manage filtration and storage will determine whether your system keeps you safe or puts you in danger.

Water can contain invisible hazards, regardless of how pure it appears. Clear, odorless water is typically a hiding place for bacteria, viruses, parasites, and chemical pollutants. Even from your own well or rain tank, drinking contaminated water can cause significant health problems. For this reason, installing a trustworthy filtration system is essential. Making your off-grid water genuinely useful requires this crucial step.

However, storage is as vital. Improper storage can make even the cleanest water in the world dangerous. Algae and bacteria thrive in stagnant water, sunlight exposure, or subpar containers. Your hard-earned water will remain safe for weeks or even months if you choose the correct tanks, maintain them sealed, and guard against contamination.

This chapter will cover the best water filtering methods, from portable alternatives to ceramic systems and easy do-it-yourself sand filters. Additionally, you will learn how to properly store water, shield it from the weather, and prepare a supply for everyday usage or long-term situations.

Life is only possible when the water is pure. Let's examine how to ensure that each drop you gather remains secure, pristine, and usable.

DIY Sand and Charcoal Filters

Purchasing pricey systems or depending on store-bought filters that eventually run out are not the only options for filtering water off the grid. With the correct supplies, you can create a straightforward yet incredibly powerful sand and charcoal filter that removes debris, grime, and some bacteria, making murky water much safer to drink.

One of the first techniques for purifying water is this type of filter. To remove contaminants and trap sediment, it slowly pushes water through layers of natural materials. A sand and charcoal filter is an excellent first step in producing cleaner water for irrigation, washing, or even livestock, but you still need to boil or treat the water afterward if you intend to drink it.

What You Will Need

The beauty of this filter is that most of what you need is simple and easy to find:

- **A large container** — a food-grade bucket, barrel, or even a tall PVC pipe works
- **Fine, clean sand** — this is your main filtering material
- Activated charcoal or hardwood charcoal crushed into small pieces
- A layer of gravel or small stones
- A piece of cloth or fine mesh
- A spigot or small hole at the bottom to let filtered water flow out

Building Your Filter — Step by Step

First, thoroughly clean your container. Since your water will flow through this, you want it to be as clean as possible.

Place your mesh or fabric at the very bottom. This permits water to flow while preventing sand and gravel from blocking the exit point.

Then, add a couple of inches of gravel. This prevents the finer components from packing too firmly and aids in drainage.

Place your layer of activated charcoal on top of the gravel. Use hardwood charcoal and smash it until the pieces are roughly pea-sized. Charcoal is a great way to absorb chemicals, smells, and contaminants.

Add your sand after the charcoal. Here, the sand takes care of the heavy lifting by capturing cloudiness and tiny particles from the water. Ensure the sand is clean and clear of dust or other impurities; the finer the sand, the better the filtration.

After assembling all your layers, gradually add some murky water to test your filter. This method relies on slow filtering, so don't rush it. You'll notice that the water is clearer and has less silt and debris.

Using and Maintaining the Filter

Remember that this filter is not quick. Its purpose is to treat water gradually so that each layer may do its function. That's the trade-off: quality for speed.

After each use, rinse the top layers, mainly the sand, to eliminate debris or muck accumulating. As charcoal loses its capacity to absorb impurities, you might need to replace it.

Neither viruses nor bacteria are eliminated by this technique. This water should always be boiled or treated with UV light or purification pills before consumption. However, this filter improves water quality for cleaning, washing, and cleaning your garden.

Solar Water Purification Methods

When you live off the grid, sunlight might be one of the easiest and safest ways to cleanse water and be a source of electricity. Solar water purification transforms dubious water supplies into much safer drinking water by using the sun's energy to eliminate dangerous bacteria, viruses, and parasites.

There's a certain satisfaction in producing clean water with just sunlight. It operates almost any place the sun shines, is silent, and doesn't need fuel or chemicals. Solar purification is a technique worth understanding, regardless of whether you are in an emergency or just trying to extend your off-grid system.

Solar Water Disinfection, or SODIS for short, is one of the simplest methods. It's easy: fill glass or transparent plastic bottles with water, screw on the caps, and leave them out in the sun for at least six hours. If the sky is overcast, you might have to keep them outside for up to two days. Most disease-causing organisms in water are killed by the combination of heat and UV radiation.

Surprisingly, SODIS works well for little water batches. Anyone may use it easily, without expensive equipment, and it's ideal for backup purification or emergencies. The technique is most effective when the water is clear because murky water hides the sunlight and lessens its efficiency. Hence, it is not recommended for widespread use or use in highly polluted water.

A solar still is an additional choice for more extensive requirements. Salts, heavy metals, and other impurities are left behind when water is evaporated by the sun in a solar still. After condensing on a cool surface, the water vapor drips into a sanitized container. A clear plastic sheet, a container for the used water, and a cup or pot to catch the clean condensation can construct a simple still.

Dig a shallow hole, fill it with soiled water, cover it with plastic, and use a small stone to weigh down the center. Water vapor rises, strikes the plastic, and drips into your clean cup when the sun heats the region. Although it takes time, the procedure yields distilled water that is free of most impurities.

Portable solar water purifiers, ideal for off-grid households or bug-out situations, are now available on the market. These devices combine tiny panels with UV filtration units. If the sun is shining, they generate clean water, are lightweight, and don't require chemicals.

Off-grid living and solar purification go hand in hand, which is its beauty. Once you know how it operates, it adds more security to your water system. Although it might not replace your primary filtration arrangement, it is a dependable backup, mainly if other systems malfunction or your resources run low.

Working with nature rather than against it is the goal of solar purification, as it is with everything off the grid. You can transform the sun's free energy into pure, life-saving water with some expertise. It only needs the quiet power of sunshine to accomplish what it does best; no fuel, chemicals, or noise are required.

Underground Water Storage Tanks vs. Above-Ground Systems:

Underground Water Storage Tanks

Naturally Insulated for Better Temperature Control

The earth's natural insulation is one of the main benefits of subterranean tanks. Once buried, the surrounding dirt keeps the water from freezing too soon in the winter and helps keep it cool during the hot summer months. This is quite beneficial in areas with high temperatures, where above-ground tanks could find it challenging to keep potable water available all year round. In addition to tasting better, the stored water stays colder and is shielded from abrupt temperature changes that might degrade water quality.

Protection from Sunlight Reduces Algae Growth and Evaporation

Subterranean tanks are totally protected from sunlight, significantly lowering the likelihood of algae formation. One of the primary causes of algae blooms, which can clog your system and cause the water to taste or smell foul, is sunlight. Additionally, keeping the tank covered avoids evaporation, a big problem in arid climates where every drop matters. This implies that more of the water you collect remains in your storage system, where it should be available for consumption.

Saves Space and Keeps Your Land Aesthetic

Underground tanks are ideal if you are concerned about how water storage might impact the look and layout of your land. Once installed, the system is completely hidden, allowing you to use the surface space for gardening, building, or simply keeping the landscape clean and uncluttered. This is especially valuable on smaller properties where every usable space matters.

Higher Installation Costs Due to Excavation and Equipment

The biggest drawback of underground tanks is the upfront cost. Installing one is not as simple as dropping a tank into place — it requires excavation, specialized equipment, and sometimes professional help. The ground must be stable and adequately prepared to prevent shifting, cracking, or collapse. Drainage systems may also need to be added to avoid flooding during heavy rains. All of this adds up, making underground systems a serious investment.

Difficult to Inspect and Maintain

While underground tanks stay out of sight, they are harder to access when something goes wrong. Inspecting for leaks, cracks, or clogs becomes a bigger job, often requiring digging or lifting heavy covers. Repairs, when needed, are usually more complicated and expensive than dealing with an above-ground system. Routine maintenance takes more planning and effort, which some off-grid homeowners find challenging.

Best for Long-Term, Large-Volume Water Storage

Despite the challenges, underground tanks are perfect for long-term use, and Soil Health manages the storage of large amounts of water. If you have the budget and land to support it, a buried tank gives you the ability to store thousands of gallons safely and securely. It becomes your quiet backup — out of sight, out of mind — until the day you really need it.

Above-Ground Systems

Affordable and Simple to Install

Above-ground water tanks are popular in off-grid setups because they are relatively affordable and easy to install. There is no need for digging or hiring heavy equipment. Using basic tools, most people can set up an above-ground system with some help. This makes it ideal for those working on a tight budget or who want to get their water storage up and running quickly.

Easy to Access for Inspections and Repairs

One of the best things about above-ground tanks is how easy they are to work on. Everything is within reach if you need to check the water level, tighten a fitting, or patch a leak. Regular maintenance is simpler and less time-consuming. You can catch problems early before they become major repairs, keeping your water system running smoothly.

Visible Water Levels Make Monitoring Easy

With above-ground tanks, it is easy to see how much water you have left. Some tanks are translucent or have gauges that let you monitor your supply at a glance. This visibility is a huge advantage in day-to-day off-grid life, where careful water management is critical. You will know when to collect more rainwater or adjust your usage — no guessing is required.

Prone to Algae Growth and Evaporation

The downside of sunlight exposure is that above-ground tanks are more prone to algae growth. Unless you keep your tank covered and sealed well, sunlight can trigger algae blooms that turn your water green and make it unusable for drinking. Evaporation is also a concern, especially in hot and dry climates. Water levels can drop faster than you realize if the tank is not adequately protected from the sun.

Takes Up Space and Affects Property Appearance

Another thing to consider is how much space above-ground tanks can take up. A few large barrels or tanks sitting in the open can quickly dominate your yard or homestead space. Some people do not mind, while others find it disrupts the natural look of their property. If aesthetics matter to you, placement and screening might take extra planning.

More Vulnerable to Temperature Fluctuations

Every change in the weather is felt by above-ground systems. The water becomes warm and occasionally less appetizing when tanks warm up in the summer sun. Water can freeze solid if the temperature drops sufficiently in the winter unless heating equipment or insulation is installed. This susceptibility should be considered while planning for those living in regions with harsh seasons.

DIY Project 3:
Construct a Sand and Charcoal Water Filter

Building your own sand and charcoal water filter is one of the most useful off-grid projects you can tackle. Using basic, reasonably priced components, you can build a system that purges water from dirt, debris, and other impurities, making it safer, cleaner, and clearer for daily use.

You can use this filter for more than just emergencies. Treating collected rainwater, well water, or surface water before drinking or using it around the homestead is a useful, long-term addition to any off-grid arrangement.

What You Will Need

A couple of robust food-grade buckets or barrels are required for this project: one for the filter and another for collecting clean water. You can make drainage holes with a drill. Additionally, you'll need multiple layers of natural materials, such as small stones or gravel, crushed hardwood charcoal or activated charcoal, and clean, fine sand. A clean cloth or a simple filter screen must hold the layers in place and capture fine debris.

Step 1 — Prepare the Bucket and Drill Drainage Holes

To begin, drill a little hole in the first bucket's bottom. The filtered water will exit at this point. Ensure your materials stay inside the opening when it is big enough to allow water to flow. If you want easier filtered water access, install a basic spigot here.

Cover the hole in the bucket with a piece of fabric or mesh. This initial barrier prevents the washing out of your filtering layers.

Step 2 — Create the Filtering Layers

Start by covering the very bottom of the bucket with a layer of tiny stones or gravel. This prevents the sand from blocking the outflow and aids in drainage. To ensure smooth water flow, uniformly distribute the pebbles.

After that, apply your layer of crushed hardwood charcoal or activated charcoal. Until the bits are about the size of peas, break them up. This layer of charcoal is effective because it eliminates some contaminants, absorbs smells, and enhances the water's flavor.

Add your fine, clean sand on top of the charcoal. This layer serves as your primary filter. Water slows down as it moves through the sand, trapping silt, debris, and even some microorganisms. Better filtration results from finer sand, but before using, ensure it is thoroughly cleaned and rinsed.

If you want to make a taller, more complete filter, you can repeat the layers of sand and charcoal. The filtration process gets slower but more efficient as you add more layers.

Step 3 — Assemble and Test the System

Once your layers are set, place the filter bucket over your clean water collection bucket. If you add a spigot, seal it tightly to prevent leaks.

Slowly pour your dirty or cloudy water into the top of the filter. Do not rush it — the slower the water moves through the layers, the better the filtration. Watch as the water drips out cleaner at the bottom.

If the water is cloudy, let the filter run a few times until the layers settle and the sand packs tighter. The first few gallons may carry fine particles, but once the system settles, the output will clear up.

Step 4 — Maintenance and Longevity

Your sand and charcoal filter will last a long time with simple maintenance. Remove the top layer of sand every couple of weeks or after heavy use, rinse it well, and check the charcoal. Replace the charcoal once it starts breaking down or losing its ability to absorb odors.

Keep the filter bucket covered when not in use to prevent insects or debris from getting in. Clean the cloth or mesh regularly to keep the water flowing smoothly.

Bonus: Common Water Filtration Mistakes & How to Avoid Them

Mistaking clear water for safe water — Just because it looks clean does not mean it is free of bacteria, viruses, or chemicals. Always filter and test.

Rushing the filtration process — Pouring water too quickly reduces effectiveness. Slow, steady flow gives the filter time to work properly.

Neglecting maintenance — Filters clog over time. Sand, charcoal, and screens need regular cleaning or replacement to stay effective.

Contaminating clean water after filtration — Storing filtered water in dirty containers or exposing it undoes all your work. Always use clean, sealed, food-grade containers.

Relying on only one system — If your filter fails, you are stuck. Always have a backup method like boiling, UV, or purification tablets ready.

End-of-Chapter 3 Checklist: Your Off-Grid Water Plan

- ☑ Water Sources Secured — Identify your primary sources (well, rainwater, surface water) and know your backups.
- ☑ Collection System Ready — Gutters are clean, screens are in place, storage tanks are appropriately sized, and overflow is managed.
- ☑ Filtration System in Place — Choose the right filters for your water type (sand, charcoal, ceramic) and know how to maintain them.
- ☑ Safe Storage Available — Store clean water in sealed, food-grade containers protected from sunlight and pests.
- ☑ Backup Purification Ready — Have a second method available (boiling, UV purifier, chemical tablets) if your main system fails.

Section 2:
OFF-GRID ENERGY INDEPENDENCE

Living off the grid entails managing energy, sourcing water, and cultivating food. No matter what happens in the outside world, power keeps your lights on, your tools operating, and your house cozy.

We explore the core of off-grid energy independence in this part. You'll learn about renewable energy sources with advantages and disadvantages, such as solar, wind, and micro-hydro systems. Along with clever strategies for long-term energy storage, we'll review alternative fuels that can keep you going when the sun or wind slows down.

We will also discuss EMP protection, including what it is, why it is important, and how to safeguard your vital systems from possible harm because the modern world is entirely of contemporary risks.

Being able to build, maintain, and safeguard your own power when you need it is what it means to be truly independent. Let's begin.

Chapter 4: Assessing Energy Needs & Reducing Consumption

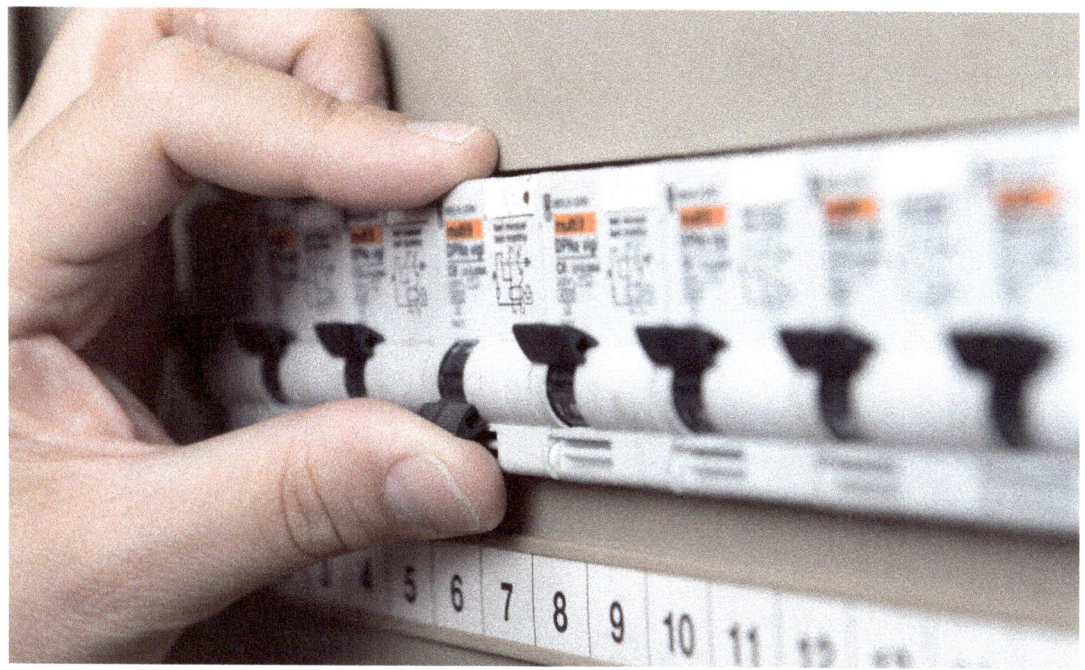

Before diving into solar panels, wind turbines, or any other energy system, the first step in off-grid living is understanding exactly how much power you need. Many people overestimate their energy requirements, leading to oversized, expensive systems. Others underestimate and end up short when it matters most.

This chapter will explain how to calculate your power needs and explore practical ways to reduce energy consumption without sacrificing comfort. The goal is simple — build a system that fits your lifestyle, not a grid-sized setup you will never fully use.

How to Calculate Your Power Requirements

Knowing your power needs starts with listing every appliance, tool, and device you plan to use — from lights and refrigerators to pumps and power tools. Each one draws a specific amount of electricity, usually measured in watts.

Start by checking the wattage label on each item. You can look it up online or use a wattmeter to measure it directly if it is not listed. Then, think about how many hours per day you use that item. Multiply the watts by the number of hours to get the daily energy use in watt-hours.

For example, if a 60-watt lightbulb runs for five hours daily, that is 300 watt-hours daily. A small fridge might draw 100 watts but run 24 hours daily, totaling 2,400 watts. Add up all your daily energy needs to find your total daily consumption.

Once you have this number, you can start designing a system that meets your actual usage rather than guessing. Remember that These are places where you want to double-check your calculations and explore alternatives.

Understanding your daily usage helps you correctly size your batteries and solar array. It also gives you a clearer picture of when you might face shortages — like during cloudy weeks or long winters — so you can plan backups and storage.

Low-Energy Living Strategies: Passive Cooling, LED Usage, and More

Living off-grid does not mean giving up comfort — it just means learning to use energy wisely. Reducing energy demand is the easiest way to stretch your power system and avoid costly upgrades.

Passive Cooling and Heating

Instead of running fans or air conditioners, passive cooling strategies can keep your home comfortable. Position windows for cross-ventilation. Use reflective roofing or install awnings to block direct sun. Thick insulation, earthen walls, or planting trees strategically around your home can stabilize temperatures without drawing a single watt.

In colder months, maximize natural heating by facing windows toward the sun and using thermal mass like stone or concrete to store heat during the day and release it at night.

Lighting: Switch to LED

Lighting is one of the easiest places to cut back on energy use. Swapping old incandescent bulbs for LED lights reduces your power draw drastically. LEDs use a fraction of the energy and last much longer. Motion sensors, solar garden lights, and dimmer switches also help you only use what you need when needed.

Efficient Appliances and Smart Usage

Choose appliances designed for off-grid living or rated for energy efficiency. DC appliances work directly with solar systems and waste less power. Use pressure cookers propane for cooking, and limit the use of large power tools to times when your battery bank is full or the sun is shining bright.

Unplug devices when they are not in use. Phantom loads — the tiny amounts of power some devices draw even when off — add up over time. Smart strips or simply pulling the plug saves power without changing your routine.

Reducing consumption is about building habits and making smart choices. Every watt you save is one less, and you must generate, store, and worry about it later. By combining accurate energy calculations with low-energy strategies, you create a system that works — simple, efficient, and reliable.

DIY Project 4:
Build an Energy Usage Tracker

Before building your off-grid energy system, getting a clear picture of how much power you use is wise. Many people either guess too high or too low — leading to wasted money or an undersized system that leaves them short. This simple DIY project helps you track your energy usage in real-time, giving you the knowledge to design a system that fits your needs perfectly.

What You'll Need

For this project, you only need a few simple tools: a notepad or notebook to log your data, a basic watt meter to measure individual appliances, and your attention to detail. Watt meters are inexpensive and easy to use — you plug your device into the meter, and it tells you exactly how much power it uses while running.

Step 1 — Make a List of Everything That Uses Power

Start by walking through your home, cabin, or campsite and writing down every electrical device or appliance you plan to use. Include lights, refrigerators, pumps, tools, chargers, fans — anything that pulls power.

Think about seasonal items, too. Maybe you only run a heater in the winter or an air conditioner in summer. Add those to the list, noting how often you expect to use them.

Step 2 — Measure Power Usage with the Watt Meter

One by one, every gadget or appliance is connected to the watt meter. Note the wattage that you observe. Run the equipment for a period of time to achieve an average reading if it turns on and off frequently, like a refrigerator. For non-plug-in devices, such as DC systems or hardwired lights, search for the wattage on the label or, if necessary, search online.

Step 3 — Calculate Daily Usage

For each item on your list, multiply the wattage by the hours you use it daily. For example:

- LED light: 10 watts x 5 hours = 50 watt-hours per day
- Laptop: 60 watts x 4 hours = 240 watt-hours per day
- Refrigerator: 100 watts x 24 hours = 2,400 watt-hours per day

Do this for every device, then add them to find your total daily power usage in watt-hours.

Step 4 — Analyze and Identify Energy Hogs

Once you have your total, review your list. See which appliances or devices are using the most power. This is where you start thinking about cutting back or replacing high-energy items with more efficient versions.

You might find that your fridge, water pump, or lighting system pulls more power than you expected. You can reduce usage by upgrading to efficient models, limiting run time, or switching to alternatives like propane or solar lights.

Step 5 — Use Your Tracker as a Planning Tool

Keep this tracker handy as you design your off-grid energy system. It gives you a clear roadmap for sizing your solar panels, battery bank, or generator. You'll know exactly how much power you need each day — no guesswork, no surprises.

Update your tracker anytime you add new appliances or change your setup. It becomes a living tool that helps you manage your energy, avoid overload, and make smart choices for the long haul.

Chapter 5: Solar Energy for Off-Grid Power

Solar energy is the main component of the power system for most off-grid residents. Once installed, it provides free electricity directly from the sun and is dependable and renewable. Solar energy helps produce a consistent flow of energy without a generator's noise or fuel requirements, whether you're using it to power equipment, charge batteries, or run lights.

The fundamentals of solar panel operation, the components of a full system, and how to appropriately size it for your off-grid requirements will all be covered in this chapter. Knowing the technology enables you to create a reliable, effective, and powerful system all year.

How Solar Panels Work

When broken down, the operation of solar panels is relatively simple despite their somewhat complicated appearance. Solar panels comprise numerous tiny photovoltaic (PV) cells, typically silicon. The purpose of these cells is to convert sunlight into electrical energy.

The electrons in the photovoltaic cells move when sunlight strikes the panel, producing direct current (DC) electricity. Imagine those electrons being shaken loose by the light and then flowing via wires attached to the panel.

Since the electricity generated is DC, it only flows in one direction. However, alternating current (AC) powers most home tools and appliances. This is the inverter's role, transforming the DC power from your solar panels or battery bank into useful AC power for your home's appliances, including lights, refrigerators, and tools.

A charge controller controls the power flow from the panels to your batteries. It shields your batteries from harm, avoids overcharging, and maintains system performance.

Therefore, solar panels produce power if the sun shines, even overcast days. You may use this power straight into your house or store it in batteries for later use. The device operates silently without fuel or moving parts, converting sunshine into reliable energy daily.

DIY Solar Power Setup

Installing your own solar power system is one of the best off-grid tasks. Once installed, the sun takes care of most of the work, producing electricity in the background as you go about your daily activities. The good news is that you can build a simple solar system that powers equipment, charges gadgets, runs lights, and even keeps a refrigerator running without being an electrician.

Here's how to create a basic solar power system that you can grow and maintain over time.

Step 1 — Gather Your Components

A basic solar power system includes a few essential parts:

- **Solar Panels** -collect sunlight and turn it into direct current (DC) electricity. Choose the size and number of panels based on how much power you need.
- **Charge Controller** — Protects your batteries from overcharging and regulates the power flow.
- **Battery Bank** — Stores your power so you can use it when the sun is not shining, especially at night or during cloudy days.
- Inverter — Converts DC power stored in your batteries into alternating current (AC) so you can run common appliances.
- **Wires, connectors, and mounting hardware** — Bring it all together and keep everything secure.

Choose deep-cycle batteries — like AGM or lithium — designed for solar storage. Regular car batteries will not hold up to daily charging and discharging.

Step 2 — Calculate Your Energy Needs

Before you buy anything, figure out how much power you really need. Add up the wattage of everything you plan to run — lights, fridge, water pump, tools — and estimate how many hours per day you will use each.

This gives you a daily total in watt-hours, which helps you size your panels and batteries properly. A small setup for lights and charging phones might only need 300 to 500 watts of solar. An entire household system may need 2,000 watts or more.

Step 3 — Mount Your Solar Panels

Choose a location with full sun exposure most of the day — rooftops, ground mounts, or even pole-mounted systems work. Ensure the panels face true south (in the northern hemisphere) or true north (in the southern hemisphere) to capture maximum sunlight.

Tilt the panels based on your latitude for the best year-round performance. Secure everything tightly so it can handle wind, rain, or snow.

Step 4 — Wire the System Safely

First, connect the charge controller to the solar panels. This stops electricity from entering your batteries before the system is prepared. To avoid overheating or voltage dips, choose the appropriate gauge wire for the size of your system.

Connect your cables to the battery bank from the charge controller. Verify your connections again, making sure they are favorable to positive and negative to negative. If necessary, make sure your batteries are adequately ventilated.

Next, attach the battery bank to the inverter. For you to operate standard household appliances, this is what transforms stored DC electricity into AC. Additionally, some inverters have USB connections or built-in outlets, which make immediate plugging in simple.

Step 5 — Test and Monitor Your System

Test your system on a bright day when everything is linked. Keep an eye on the charge controller display to watch the batteries charging and the panels generating power. To determine how much electricity you use, try running a few appliances and checking the inverter.

Pay attention to your battery levels, particularly in the beginning. You will eventually learn when to save energy and when to use more.

Step 6 — Maintain Regularly

Solar systems are low maintenance, but they do need regular checks. Clean your panels every few months to keep them efficient — even a layer of dust can cut performance.

Inspect wires, connections, and batteries monthly. Look for corrosion or wear. Ensure your charge controller settings match your battery type and are working correctly.

Of course! Here's a **natural, easy-to-read comparison** of Battery **Storage: Lead-Acid vs. Lithium-Ion,** expanded into entire paragraphs for clarity:

Battery Storage: Lead-Acid vs. Lithium-Ion

When it comes to off-grid solar power, your battery bank is the heart of the system — storing the energy you collect so you can use it when the sun is down. Choosing the right type of battery can make a huge difference in how efficient, reliable, and cost-effective your setup is over time. The two most common options are Lead-Acid and Lithium-Ion batteries, each with strengths and drawbacks.

Lead-Acid Batteries — Reliable but Heavy and Maintenance-Heavy

Lead-acid batteries have been the conventional option for off-grid devices for many years. They are typically the least expensive upfront and have been tried and proven. Deep-cycle lead-acid batteries, such as flooded or AGM models, are a good choice for solar storage since they can withstand frequent charging and draining.

The availability of lead-acid batteries is one of its main benefits. They are comparatively simple to set up and can be found practically anywhere. They are frequently the least expensive option to get your system up and running if money is an issue.

Lead-acid batteries can have certain drawbacks, though. If you have limited space, you should consider that they are big, hefty, and occupy much room. Additionally, they need to be maintained regularly, particularly if you're employing flooded versions. You must clean terminals, check water levels, and monitor their condition to prevent damage.

Their usable capacity is another crucial component. Discharging lead-acid batteries below 50% of their capacity is not recommended without reducing their lifespan. This implies that to satisfy your daily needs, you will need a battery bank that is significantly larger than you may initially believe. Additionally, with proper maintenance, they usually survive five to seven years.

Solar Energy for Off-Grid Power

Lithium-Ion Batteries — Expensive but Efficient and Low-Maintenance

With good reason, lithium-ion batteries are quickly taking the lead in off-grid solar systems. They are highly efficient, lighter, and more compact. Discharging a lithium battery to 80 or even 90 percent of its capacity without causing damage or reducing its lifespan is possible. This implies that a smaller battery bank can provide more helpful power.

Their longevity is yet another significant benefit. A high-quality lithium-ion battery bank can endure for ten years or longer with little care. Lithium batteries are sealed and safe to install indoors, so there's no need to top off water levels or worry about ventilation.

Additionally, they charge more quickly than lead-acid batteries, maximizing your solar harvest on overcast or brief winter days. Lithium is challenging to match if space, effectiveness, and long-term value are important considerations.

The most significant drawback? Price. The initial cost of lithium-ion batteries is still much higher. However, many off-grid homes' improved performance and longer lifespans eventually make the investment worthwhile. You save space as well as upkeep and replacements.

Common Solar Panel Myths & Mistakes

1. Myth: Solar Panels Work Only on Sunny Days

One of the most widespread myths is that wet or overcast days cause solar panels to malfunction. In actuality, solar panels continue to produce electricity during cloudy skies but with decreased efficiency. Even though the greatest amount of electricity is produced under direct sunlight, your system will still gather energy whenever there is sunshine. Effective solar systems are made to withstand different types of weather and maintain battery charge throughout the year.

2. Mistake: Oversizing or Under sizing the System

Many people estimate their energy requirements and either build an inadequate system or overpay for too many panels. You pay for both errors. While undersizing causes annoyance when your batteries run out too soon, oversizing loses money. The secret is figuring out how much energy you consume daily, appropriately designing your battery bank and solar array, and factoring in extra loads or overcast days.

3. Myth: Solar Panels Are Maintenance-Free

It is easy to assume that once your panels are installed, you can forget about them. While they are low-maintenance, they are not maintenance-free. Dirt, dust, snow, and leaves reduce efficiency if left too long. Cleaning your panels a few times a year, checking wiring for wear, and tightening loose connections ensures your system stays efficient and lasts longer.

4. Mistake: Poor Panel Placement

A solar system's placement makes or ruins it. Panels must be oriented correctly according to your latitude and face the correct direction, typically true south in the northern hemisphere. The output of your system is significantly reduced if you mount them where trees, chimneys, or other structures provide shadow during peak hours. Before installation, always survey your property and measure the hours of sunlight.

5. Myth: Solar Systems Are Too Expensive to Be Worth It

Although solar power was once costly, costs have significantly decreased during the past ten years. Most people can now afford tiny off-grid systems, especially compared to the long-term costs of electricity bills or operating generators. With proper design, you may create an independent system that saves gasoline and pays for itself in a matter of years.

6. Mistake: Neglecting the Battery Bank

Many people focus entirely on panels and forget that the battery bank is just as important. Cheap or undersized batteries will wear out fast, leaving you without power when the sun goes down. A well-sized, high-quality battery system — lead-acid or lithium — keeps your power flowing day and night, storing energy when needed.

7. Myth: Solar Power Alone Can Handle All Energy Needs

Solar is powerful, but it is not magic. Depending on your location, weather, and energy habits, there will be times when solar alone might not be enough — especially in winter or during long, cloudy stretches. Smart off-grid setups often include backup systems like wind, micro-hydro, or a small generator to cover gaps and avoid running out of power.

8. Mistake: Forgetting About Inverter Sizing

Your inverter turns the DC power from your panels and batteries into AC power for household use. Too many people skimp on inverter size only to discover they cannot run their fridge, power tools, or pump without overloading the system. Choosing the right inverter size — based on peak loads — is crucial for smooth, reliable operation.

9. Myth: Adding More Panels Solves Every Power Problem

It is easy to assume that you just need more panels if you run low on power. But your system is only as strong as its weakest part. If your battery bank is small or your charge controller cannot handle the extra input, adding panels won't fix the problem — it might even damage your system. Balance is key — solar panels, batteries, charge controllers, and inverters must all be sized to work together.

10. Mistake: Ignoring Local Weather and Seasonal Changes

Some folks build their solar system based on the summer sun — and then find themselves short on power when winter hits. Failing to account for shorter days, lower sun angles, and snow can drain your batteries and your home without power. Always design your system for the toughest season, not the easiest. That way, you are ready year-round.

DIY Project 5:
Install a Small Solar Panel System

Installing your own small solar panel system is one of the most rewarding off-grid projects you can tackle. Whether powering a cabin, shed, or camper or simply creating backup power, this system gives you reliable energy from the sun.

With essential tools, planning, and patience, you can wire and install a simple solar setup that covers lights, small appliances, and essential devices.

Tools & Materials You'll Need

- Solar panels (sized to your energy needs)
- Deep-cycle batteries or lithium battery bank
- Solar charge controller
- Inverter (sized to run your appliances)
- Wires, connectors, fuses, mounting brackets, and basic hand tools

Step 1 — Plan and Size Your System

Before you start, calculate how much power you need daily. Add the wattage of lights, devices, and appliances you plan to run. Choose solar panels that produce enough watt-hours daily to meet your needs, plus a little extra for cloudy days.

Determine battery capacity based on your usage and how many days of stored power you want available. Remember to size your inverter based on the largest load you plan to run.

Step 2 — Set Up Your Solar Panels

Pick a sunny location — rooftops, ground mounts, or poles work well. Mount your solar panels securely using brackets or racking systems angled to capture maximum sunlight. Make sure they face the true south (in the Northern Hemisphere) or true north (in the Southern Hemisphere).

Check that the panels are tilted correctly for your region and that nothing shades them during peak hours.

Step 3 — Install the Charge Controller

Connect the solar panels to your solar charge controller first. The charge controller protects your batteries from overcharging and helps regulate the flow of electricity.

Always follow the positive-to-positive and negative-to-negative rules when connecting wires. Double-check the manual for your specific controller model to ensure correct connections.

Step 4 — Wire in the Battery Bank

Run heavy-gauge wire from the charge controller to your battery bank. Install fuses or breakers for safety. Ensure all terminals are clean and tight — bad connections can drain your system or cause overheating.

If you are using lead-acid batteries, ensure the area is well-ventilated. Lithium batteries can usually be placed indoors safely.

Step 5 — Connect the Inverter

The inverter turns your stored DC power into AC electricity so you can run regular household appliances. Wire the inverter directly to the battery bank, keeping the wires short and thick to handle the power draw.

Once connected, test the inverter by plugging in a small device — like a light or fan — to check that it works properly.

Step 6 — Test Your System

Monitor the charge controller to ensure the panels generate power on a sunny day. Verify that the inverter is operating your gadgets smoothly and observe how your battery levels increase.

Verify that the panels, batteries, and inverter are operating as intended by testing each system component.

Step 7 — Maintain Regularly

Check wiring and connections monthly. Clean the solar panels to remove dust, dirt, or snow buildup. Monitor battery health and replace worn parts before they fail.

Chapter 6: Wind & Hybrid Energy Solutions

Wind energy is another potent instrument for attaining complete energy independence, even if solar power receives the majority of the focus on off-grid living. The wind produces power when it blows, regardless of the time of day or night. Installing modest wind turbines in areas with consistent winds can help keep your batteries charged at night, on overcast days, or in the winter when the sun isn't shining as brightly.

Even better is a hybrid system that balances the advantages and disadvantages of solar and wind power. You may capture solar energy when the sun is shining. Your wind turbine takes over when the weather gets stormy or the seasons change.

This section will teach you how to install a basic wind turbine system, which is achievable, realistic, and intended to give your off-grid power arrangement additional robustness.

Setting Up a Small Wind Turbine (DIY)

Adding a small wind turbine to your energy system is a great way to keep power flowing when the sun is not enough. It is not as complicated as you might think — with the right tools, materials, and location, you can install one over a weekend.

Step 1 — Choose the Right Spot

Location is the most crucial component of any wind arrangement. Clean, consistent wind is necessary for your turbine to produce electricity effectively. The best places include hills, ridgelines, broad fields, and places far from towering trees or structures. Because wind speeds rise with height, the higher your turbine, the better.

A tower should be at least 20 to 30 feet above any surrounding obstructions. Lifting the turbine can have a significant impact if ground-level wind is erratic.

Step 2 — Gather Your Materials

For a basic system, you will need:

- A small wind turbine rated for your power needs (300 to 1000 watts is common for off-grid setups)
- Tower or pole for mounting
- Guy wires and anchors for stability
- Charge controller made for wind systems
- Battery bank (or connect to your existing solar battery system)
- Inverter (if running AC appliances)
- Heavy-gauge wiring, connectors, and safety fuses

Many small wind turbines come as kits with the blades, hub, and generator ready to assemble.

Step 3 — Install the Tower and Turbine

Assemble your tower or pole first. If you use a guyed tower, anchor your wires securely to prevent tipping in strong winds. Safety here is critical — wind turbines create a lot of force once they start spinning.

Mount the turbine at the top of the tower, following the manufacturer's instructions. Make sure the blades spin freely and are balanced. Poor balance causes vibration and shortens the lifespan of your system.

Run your heavy-duty wire down the tower, securing it to prevent damage from movement or weather.

Step 4 — Connect to the Charge Controller and Batteries

Even when your batteries are fully charged, wind turbines produce electricity anytime the wind blows. A wind-specific charge controller is therefore required. It controls the flow of power and guards against overcharging or battery bank damage.

Run the output to your battery system after connecting your turbine cabling to the charge controller. You can create a genuine hybrid system by allowing the wind turbine to share storage with your existing solar battery bank.

Step 5 — Monitor and Fine-Tune

Once installed, monitor your system closely for the first few days. Watch how your batteries charge and track the power output during windy periods. Adjust your guy wires if needed, and ensure all connections are secure.

Some wind systems include a dump load — a resistor or heater that safely burns off extra power when your batteries are full. Ensure it is installed and working to avoid stressing the system during high winds.

Combining Solar and Wind for Maximum Efficiency

Relying solely on one power source can expose you to risks regarding off-grid electricity. Clear, sunny days are ideal for solar power, but lengthy winter nights or overcast conditions can deplete your batteries more quickly than the sun can replenish them. In contrast, wind energy increases during colder months when sunshine is less abundant or when storms move in.

For this reason, one of the most well-rounded and effective off-grid living configurations is produced by integrating solar and wind into a hybrid system. These two natural forces work together to fill in the blanks and ensure a consistent power supply throughout the year.

Long summer days with strong, high sun are ideal for solar panels. At this point, your system produces more electricity than you usually require. However, that consistent solar supply may decrease as fall approaches and the days grow shorter if a week of overcast weather sets in. Wind power frequently excels at this precise moment. While many places have more regular wind patterns in the winter, stormy weather produces greater winds when solar generation is at its lowest.

You can generate a dynamic energy source by connecting both systems to the same battery bank. In the summer, your solar panels can manage most of the demand, and your wind turbine takes over in the windier, colder months. A constant breeze keeps your batteries charged, even when the sun isn't shining as brightly.

Resilience is another benefit of hybrid systems. You still have a backup power source in case one of your systems fails, such as your solar inverter malfunctioning or a panel being damaged by a storm; in an off-grid scenario where every bit of electricity matters, that redundancy is invaluable.

Large battery banks are also less necessary with hybrid configurations. After a day of intensive use, your system recovers more quickly since you're charging from two sources. Enabling you to size your storage rather than overbuilding to handle the worst-case solar-only scenarios can save you money upfront.

Balance is the key to making this work. Both systems must be sized according to your environment and needs. You may rely more on your turbine if you reside in an area with consistent, high winds. More of the demand will be supported by solar power if the wind is weak and erratic. Mixing the two gives you versatility; even in inclement weather, flexibility keeps you comfortable.

The freedom and peace of mind that come with true energy independence are ultimately yours when you transform your off-grid homestead into a self-sufficient powerhouse with a solar and wind hybrid system. This system operates day and night throughout the year.

Hydroelectric Micro-Generators for Those Near Water

One of the most dependable off-grid power sources is moving water, which you can use if you are fortunate enough to have a creek, stream, or small river through your land. Water flows day and night, frequently with amazing constancy, in contrast to the sun or wind, which depend on the weather. This is the role of hydroelectric micro-generators, compact, effective devices that convert the continuous flow into reliable, sustainable electricity.

Micro-hydro systems direct a section of your river or stream via a penstock, a pipe that transports the water downhill to a turbine. The turbine is rotated by the rushing water, producing electricity that may be utilized immediately or supplied straight to your battery bank. When your system is set up correctly, the water is discharged back into the stream after being powered, leaving no environmental impact.

The fact that micro-hydro power generates energy continuously is its true charm. A well-designed hydro system can provide electricity around the clock as long as the water flows, while solar panels sleep at night and wind turbines wait for a breeze. Because of this, it is among the most reliable energy sources for off-grid life, particularly in areas with mountains or forests and year-round water supplies.

Efficiency is an additional benefit. If your site has good water flow and elevation drop (sometimes called "head"), micro-hydro systems can frequently provide more power per dollar than solar or wind. Over time, even a modest system can deliver hundreds of watts consistently, significantly more helpful electricity than a solar array of the same size.

Of course, hydro has its share of problems, just like any other system. The site must have a consistent, year-round flow and sufficient head to generate pressure. If the stream slows in the summer or freezes in the winter, seasonal variations may result in lower productivity. Additionally, the upfront work of establishing the turbine, penstock, and intake typically entails pipe laying and excavation.

However, a micro-hydro system is long-lasting and requires little maintenance once it is operating. The system can function for years with little maintenance if properly screened to keep debris out. Additionally, since your system is constantly recharging, you can frequently utilize fewer batteries because the energy production is so consistent.

A micro-hydro generator can silently power a house, workshop, or small farm for those living near flowing water. It transforms your land's natural flow into a reliable, sustainable energy source that will continue functioning during the changing seasons, while you're working, and while you sleep. It is an off-grid power solution that genuinely works with nature, not against it, and is definitely worth investigating if you have access to water.

DIY Project 6:
Build a Simple Wind Turbine

Building your own wind turbine is an excellent approach to expanding your off-grid system's renewable energy sources. While huge commercial turbines are costly, a modest, do-it-yourself turbine can quickly produce enough electricity to run lights, charge batteries, and power vital systems, especially when the sun isn't shining.

You can build a small wind turbine that harnesses the energy of the breeze and converts it into useful electricity with simple tools, supplies, and a little effort.

Tools & Materials You'll Need

- PVC pipes (for blades — 4 to 6 inches diameter, 24 inches long)
- Permanent magnet motor or DC motor (rated 12V-24V)
- Charge controller (wind compatible)
- Metal or wooden hub for mounting blades
- Mounting pole or pipe (10–20 feet tall)
- Drill, saw, wrenches, bolts, and screws
- Heavy-gauge wire and connectors
- Bearings or bushings (optional for smoother rotation)

Step 1 — Build the Blades

Start by cutting your PVC pipe lengthwise into three or four curved blades — each about 24 inches long. Taper the edges so they catch the wind efficiently. You want the wide part facing into the wind and the narrow part trailing behind.

Sand the edges smoothly to reduce drag and improve spin. The smoother the blades, the better the turbine performs.

Step 2 — Prepare the Motor and Hub

Attach the blades evenly around your hub — this could be a metal disc or a thick wooden piece drilled to fit the motor shaft. Balance is key here. Uneven blades cause vibration and waste energy.

Secure the blades with bolts and double-check that they spin freely without wobbling. Mount the hub onto your motor shaft tightly so it turns smoothly as the blades spin.

Step 3 — Mount the Motor on a Pole

Choose a tall pole or pipe for your tower — 10 to 20 feet is a good start. The higher the turbine, the stronger and steadier the wind.

Attach the motor securely at the top of the pole, ensuring the blades can spin freely in any direction the wind blows. You can create a simple tail vane from sheet metal or wood to help the turbine face into the wind.

Step 4 — Wire the System

Run a heavy-gauge wire from the motor down the pole to your wind charge controller. The controller protects your battery bank from overcharging and manages power flow.

Connect the output from the charge controller to your battery bank or storage system. Add fuses or breakers for safety.

If you want to power AC appliances, connect your battery bank to an inverter — just like you would with solar.

Step 5 — Test Your Wind Turbine

Once everything is wired and mounted, wait for a breezy day and watch your turbine spin. A multimeter is used to measure the voltage coming from the motor. You should see power flowing when the blades turn.

Monitor your battery levels and check that the charge controller is regulating properly. Adjust the blade angle or tail if needed for better wind capture.

Step 6 — Regular Maintenance

Like any off-grid system, your homemade wind turbine will need regular checks. Inspect bolts, wiring, and blades every month. Clean the blades and oil any moving parts to keep everything spinning smoothly.

Monitor your battery levels, especially during strong winds when your turbine generates more power than expected.

End-of-Chapter Checklist: Building Your Off-Grid Energy System

- ☑ Energy Needs Calculated — Have you measured your daily energy use (in watt-hours) and sized your system based on actual numbers, not guesses?
- ☑ Solar System Planned or Installed — Is your solar array sized correctly, mounted in full sun, and connected to a charge controller and battery bank?
- ☑ Wind Power Considered or Added — If you live in a windy area, have you installed or planned for a wind turbine to complement your solar setup?
- ☑ Battery Bank Sized and Ready — Have you chosen the correct type of batteries (lead-acid or lithium) and sized your storage to cover days with little sun or wind?
- ☑ Inverter Installed — Is your inverter powerful enough to handle your peak energy loads for appliances, tools, and devices?
- ☑ Backup Plan in Place — Do you have a generator or an alternative energy backup for emergencies or extended periods of low sun and wind?
- ☑ Energy-Saving Strategies Active — Have you incorporated low-energy appliances, LED lighting, passive heating and cooling, and good energy habits?
- ☑ Maintenance Plan Ready — Are you prepared to regularly check panels, turbines, batteries, and wiring to keep your system running smoothly?
- ☑ Hybrid System (Optional but Recommended): Have you considered combining solar, wind, and possibly hydro for a balanced, resilient system?
- ☑ Future Expansion Planned — Is your system designed with room to grow if your energy needs change later?

Section 3:
FOOD SECURITY & SUSTAINABLE LIVING

Learning to rely on your land, your own hands, and the cycles of nature to keep your pantry stocked is what it means to live off the grid. Creating a system where you have control over what ends up on your table, season after season, without relying on shaky supply chains or overcrowded stores, is what food security is all about.

This part examines the true meaning of growing your own food, from raising livestock that generates revenue to growing gardens that flourish in your climate. You'll discover how to stretch your resources, preserve the harvest, and establish a dependable food supply to sustain your family through prosperous and difficult times.

Finding methods to give back to the land as much as you take, creating soil rather than removing it, and collaborating with the natural cycles in your environment are all components of sustainable living. Every action to improve your self-sufficiency is cultivating heirloom seeds, composting leftovers, or learning how to preserve food.

The straightforward objective is to live so that your next meal depends on you, your land, and your abilities. The resources, information, and motivation you need to transform your off-grid house into a fully self-sufficient homestead are in the following pages.

Chapter 7: Off-Grid Cooking Essentials & Fuel Sources

Cooking off the grid necessitates a mental change because you aren't just pressing a button or switching a switch. Instead, you use natural resources like fire, sunlight, and creativity to cook your food. The good news is that cooking without electricity gives you access to various valuable instruments and traditional techniques that not only complete tasks but also help you rediscover the fundamentals of self-sufficiency.

This chapter examines the necessary off-grid cooking equipment every farmhouse needs and reliable, sustainable fuel sources when resources are limited or the electricity goes out. With the correct setup, cooking becomes necessary and a skill worth mastering, whether baking bread in the sun or boiling stew over an open flame.

Cooking Without Electricity (Rocket Stoves, Solar Ovens, Fire Pits)

Being able to cook without power is a huge advantage while living off the grid. Alternative cooking techniques are crucial for power outages, fuel shortages, or the need to live simply. Thankfully, there are tried-and-true methods for making sustainable and effective meals without electricity.

Rocket stoves are one of the most efficient cooking tools you can build or buy for off-grid living. With just a tiny quantity of wood, twigs, or other dry biomass, their straightforward design—a vertical combustion chamber with a horizontal fuel feed—produces very high temperatures. The design makes the fire burn cleaner and hotter, which reduces smoke and speeds up cooking. A rocket stove is ideal for boiling water, frying, or simmering because it uses little fuel and produces food rapidly, which is very helpful when wood is scarce.

Solar ovens harness the sun's energy, turning sunlight into heat that slow-cooks meals without fuel. To achieve temperatures high enough to bake bread, roast meat, or boil a pot of beans, these straightforward but ingenious arrangements use reflective panels to trap heat inside an insulated box. Solar ovens are an excellent low-effort cooking choice because they operate best on bright, sunny days and require no maintenance once your food is inside. The best part is that they are entirely free to run once constructed.

Fire pits remain the oldest and most reliable way to cook outdoors. Whether digging a simple pit or building a stone ring, cooking over open flames connects you to the basics. You can grill, bake in Dutch ovens, or hang pots over the flames for stews and soups. Fire pits double as a heat source and gathering place, especially during colder months. The secret is to master fire management, which involves regulating cooking durations and heat levels. This takes work but yields benefits in terms of adaptability.

There are benefits to each of these methods, including rocket stoves, sun ovens, and fire pits. They give you options regardless of the weather, season, or fuel source. If you get adept with them, you won't ever need to rely on the grid to provide a hot meal.

DIY Smokers & Dehydrators for Food Preservation

Knowing how to preserve the food you cultivate, hunt, or grow is one of the most crucial off-grid living abilities. You need conventional techniques that enable you to preserve food securely for months, even during difficult times or seasons, if you don't have a refrigerator or freezer that runs constantly. Two of the oldest and most dependable preservation methods are smoking and drying, and the greatest news is that you can make your own smoker or dehydrator at home.

DIY Smokers — Preserving with Smoke and Heat

In addition to adding flavor, smoking is a tried-and-true method of preserving meat, fish, and even some vegetables. When food is properly smoked, it dries up and absorbs natural preservatives from the wood smoke, extending its shelf life without refrigeration.

Old barrels, metal sheets, or clay bricks are simple DIY smoker-building components you probably already have. In the basic design, hardwoods like oak, apple, hickory, or maple are burned in a firebox. Your meat or fish hangs or rests on racks in a separate smoking chamber, where the smoke is directed, and the meat or fish absorbs the smoke as it gradually dries out.

A low temperature, often below 90°F, is ideal for cold smoking, which adds flavor and preserves meat without cooking. With hot smoking, the temperature is raised to the point where the meat is completely cooked while being smoked, making it ideal for immediate consumption instead of long-term storage.

Temperature and airflow management are essential for a good smoker. You may control the smoke flow by adding vents or movable flaps, and you can stay within the proper range by using a thermometer. Large quantities of fish, jerky, bacon, or game meat can be preserved with a well-made smoker, ideal for expanding your off-grid pantry.

DIY Dehydrators — Harnessing Heat and Airflow to Preserve Your Harvest

Another great method of preserving food is dehydrating it, particularly fruits, vegetables, herbs, and even thin slices of meat for jerky. Eliminating moisture stops germs, mold, and spoiling, and the finished product is small, light, and ideal for storage.

A do-it-yourself dehydrator is easy to construct. You need constant warm air to extract the moisture and racks or screens to arrange your food. One of the greatest off-grid solutions is a solar dehydrator, which requires only sunshine and patience and no electricity.

Clear plastic or glass panels, mesh screens, and wood can be used to construct a simple sun dehydrator. While vents or a chimney generate a natural flow that pulls moisture out of the food and the top, the sun heats the air inside the box. Efficiency is increased when the box is tilted toward the sun.

If the weather is uncooperative, you can construct a basic wood-fired dehydrator by channeling warm air into the drying chamber from a low, constant heat source, such as a tiny fire or rocket stove.

Dehydration's adaptability is what makes it so lovely. Apples, berries, herbs, mushrooms, greens, and meats can all be dried. The final product is lightweight, shelf-stable, and ready to use in soups, stews, or trail snacks. It stores nicely in jars, cloth bags, or sealed containers.

Off-Grid Cooking Essentials & Fuel Sources

DIY Project 7:
Build a Rocket Stove

One of the easiest and most fuel-efficient outdoor cooking devices you can construct is a rocket stove, which is ideal for off-grid life. It is perfect for boiling water, cooking food, or even heating small areas without wasting fuel because it produces tremendous heat while burning small amounts of wood, twigs, or biomass.

The finest aspect? With simple, low-cost materials, you may construct one yourself; no specialized tools are required.

Tools & Materials

- Fire bricks or regular clay bricks (around 20-30, depending on size)
- Metal tubing or stovepipe (about 4-6 inches in diameter)
- Clay or mud (for sealing gaps and insulation)
- Shovel, level, and basic hand tools

Step 1 — Choose a Safe Spot for Your Rocket Stove

Pick a flat, level outdoor area where the stove will be stable and safe. Keep it away from flammable materials, and if possible, place it under a covered area or build a simple shelter to protect it from rain.

Step 2 — Build the Base and Fire Chamber

First, build a square or rectangular foundation by creating a sturdy basis with your bricks. Make room for the fuel feed tube, where you install the metal tubing, or make a brick channel to feed kindling and sticks into the fire.

The combustion chamber, where the fire burns hottest, is created vertically stacking bricks. To use metal tubing as your fuel feed, slip it into the base horizontally. The vertical combustion chamber should draw the heat and flames as it rises above this intersection.

Step 3 — Create the Heat Riser

Use a metal pipe or brick stacking to construct your heat riser above the combustion chamber. The efficiency of the rocket stove lies in this: hot gasses from the fire shoot up this riser, concentrating heat in one area.

Pack mud or clay around the heat riser to insulate it and maintain the heat's focus for even greater efficiency.

Step 4 — Add the Cooking Surface

Once your riser is complete, create a flat, sturdy surface to place pots or pans. You can rest a metal grate on a thick sheet or simply balance your cookware on bricks spaced wide enough to let heat escape.

Ensure enough gap for air and smoke to flow upward while keeping your cooking surface stable.

Step 5 — Test and Fine-Tune

You don't need big logs; gather dry kindling, twigs, and small sticks. Light the fire after feeding them into the horizontal fuel tube. As the flames are brought into the combustion chamber, a roaring fire with little smoke should be produced.

To increase the draft and heat, modify the fuel supply or, if necessary, add more insulation.

Step 6 — Use and Maintain

Once it's burning well, you can start cooking. Rocket stoves heat up fast, so keep an eye on your food. Use small sticks as fuel and feed them slowly to maintain steady heat.

After each use, clear out ash buildup to keep airflow strong. With basic care, your rocket stove will last season after season.

Chapter 8: Food Gathering, Hunting & Preservation

Living off the grid alters your relationship with food. Walking into a store is no longer enough; you need to read the seasons, know the area, and use your talents to hunt, collect, and conserve what you need. One of the most important survival skills you may acquire is the ability to obtain sustenance directly from nature, whether it be fish, animals, or natural plants.

People have survived for millennia by hunting, fishing, and foraging for edible plants; these abilities are still crucial for anyone serious about becoming self-sufficient today. However, obtaining food is only half the problem. Knowing how to properly preserve it once you have, it guarantees that nothing is wasted and that your laboriously harvested food will last you through the seasons.

The practical skills of tracking and hunting responsibly, gathering wild foods, and storing anything from meat to berries will be covered in this chapter so that your off-grid pantry remains stocked even during times when fresh food is in short supply. Here, tradition and survival collide, and knowledge transforms nature into your most valuable asset.

Hunting & Trapping Basics

Since hunting and trapping have been used to feed families for thousands of years, these skills can help you turn the wild environment into a dependable source of protein when living off the grid. Knowing you can use skills passed down through the years to provide meat for the table with your hands gives you a profound sense of fulfillment.

Understanding your land and the game it contains is the first step. Knowing the patterns of the animals, which vary by region and include deer, rabbits, wild birds, turkeys, and even small game like squirrels, is half the fight. Learn about the routes taken by animals, their feeding habits, and the indicators they leave behind. If you learn to read, you can tell a tale from feeding spots, scat, tracks, and worn pathways.

Placement of shots and ethical harvesting are important aspects of hunting. The objective is a clean, speedy kill, whether using a rifle, shotgun, bow, or crossbow. To respect the animal and the land and for your success, practice frequently until you feel comfortable.

Trapping is helpful for people who want to hunt more passively or lay food lines while completing other duties. You can use cage traps, deadfalls, and snares to catch smaller animals like squirrels, raccoons, and rabbits. However, when it comes to traps, preparation and understanding are crucial. Since many places restrict what, where, and how you can trap, you should be aware of the rules in your area.

Placing your trap where the animal will naturally walk or pass through is essential to successful trapping. Bait is helpful, but the greatest traps are in busy areas where animals are drawn to your setup by scent and curiosity. Additionally, checking a trap every day after it has been set is morally right, keeps your catch fresh, and prevents needless suffering.

In addition to food, hunting, and trapping yield precious resources like hides, bones, and sinew that can be utilized for tools, clothing, and other necessities of life. If you take the time to acquire traditional processing skills, nothing has to be wasted.

Taking what you need, using what you take, and always showing respect for the land and the creatures that provide for you are the cornerstones of ethical hunting and trapping. When used properly, these abilities become more than survival skills; they provide a connection to nature that few other things can.

Foraging Edible Wild Plants

One of the first human abilities is foraging, which becomes a vital tool for augmenting your food source while you're living off the grid. There are many delicious and medicinal wild plants on the ground around you, but effective foraging requires knowledge, perseverance, and a profound respect for the natural world.

Knowing exactly what you're picking is the first and most basic rule of foraging. While many edible plants are growing in the wild, numerous similar species can cause illness or worse. Before you begin tasting anything new, always have a good field guide tailored to your area with you, or even better, get direct advice from knowledgeable locals or experienced foragers.

Choosing common, clearly identifiable plants is one of the simplest methods to start foraging. For instance, dandelions can be found practically anywhere. They have vitamin-rich leaves, tasty blooms, and tea-making roots that can be roasted. Blackberries, raspberries, and elderberries are examples of wild berries that are also suitable for beginners; however, be cautious to definitely identify them, as certain berries are poisonous.

Foraging is a seasonal activity. Tender greens are ideal for soups or sautés, such as wild mustard, lamb's quarters, chickweed, and nettles brought in by spring. Berries ripen, wild herbs thrive, and delicious flowers blossom as summer approaches. As fall approaches, it's time to harvest nuts like walnuts or acorns and search for roots like wild carrots or burdock.

Learning how to identify plant habitats and growing conditions makes a big difference. Common locations for edible greens include meadows, the margins of forests, and areas close to water sources. Another essential wild food is mushrooms, which appear after rain but should be handled with special care because many edible species have poisonous lookalikes.

Harvest sustainably. Always give enough space for the plant to recover and develop again, and never completely remove anything from an area. Harvesting berries and nuts sparingly is a good idea so wildlife can still have their part. Instead of destroying the land, the objective is to develop a relationship with it.

Another piece of advice is to always be mindful of pollution risks. Steer clear of foraging near industrial areas, roadsides, or locations where herbicides or pesticides may have been utilized. Although the vegetation there may appear healthy, it may be tainted.

Finally, learn how to safely prepare and process wild plants. Some are best boiled to eliminate bitterness or possible irritants, such as nettles or cattail shoots. Others can be consumed fresh, such as mint or wild garlic. You'll develop a knowledge base about which plants are best preserved, dried, or fresh over time, making foraging a consistent and dependable component of your off-grid food system.

When done correctly, foraging is more than just a means of surviving; it's a means of strengthening your ties to the land, increasing your food supply, and growing in appreciation for the environment that sustains you.

Long-Term Food Preservation: Fermentation, Canning, and Drying

Being off the grid requires planning, particularly about food. It's only half the work to grow or collect enough to stock your pantry. The true difficulty is learning how to preserve what you gather to last through the winter, dry seasons, or periods when fresh food is in short supply. Fermentation, canning, and drying are three of the most dependable and tried-and-true techniques; each offers a unique way to increase the shelf life of your food while preserving its flavor and nutrients.

Fermentation Preserving Through Good Bacteria

One of the earliest and most organic preservation techniques is fermentation. It prevents harmful bacteria from growing while fostering the ideal conditions for good bacteria, such as lactobacillus, to flourish. As these beneficial bacteria proliferate, they generate alcohol or acids that improve flavors, preserve food, and even increase its nutritional worth.

Fermented foods only need clean jars, salt, and patience; they don't need electricity or costly equipment and keep well for months.

Best foods for fermentation include:

- Cabbage (sauerkraut and kimchi)
- Cucumbers (classic pickles)
- Carrots, beets, radishes, and other crunchy vegetables
- Garlic and onions

- Dairy (yogurt, kefir)
- Fruits (small-batch fruit chutneys or preserves)
- Grains (sourdough starter)

In addition to preserving, fermentation enhances digestion and supplements your diet with probiotics, a considerable advantage when living off the land.

Canning Sealing in Freshness for the Long Haul

A reliable technique is canning, which involves sealing the food in airtight jars after using heat to destroy molds, germs, and yeast. Canning is ideal for stocking an off-grid pantry since once sealed, these jars can be stored for a year or more.

Pressure canning is used for low-acid meals, meats, and stews, while water bath canning is used for high-acid foods. All you need for water bath canning is a big kettle, jars, and lids. Although it requires a little more work, pressure canning makes it possible to properly preserve a greater variety of foods.

Best foods for canning include:

- Tomatoes, tomato sauce, salsa
- Fruits (peaches, apples, berries, pears)
- Jams, jellies, fruit preserves
- Pickled vegetables
- Soups, stews, chili
- Meats (chicken, beef, fish)
- Beans and legumes

Canning locks in flavor and nutrition, making it possible to enjoy garden-fresh produce and home-cooked meals all year, no freezer needed.

Drying Removing Moisture to Preserve Nutrients

One of the simplest and best preservation methods is drying. Eliminating moisture stops germs and mold from forming, extending the shelf life of food for months or even years if stored correctly. Dehydrated food is lighter, more portable, and ideal for long-term storage or hiking.

Food can be dried with low heat from a wood stove, oven, sun, or handmade dehydrator.

Best foods for drying include:

- Fruits (apples, bananas, berries, peaches, plums)
- Vegetables (tomatoes, peppers, zucchini, green beans)
- Herbs (mint, oregano, thyme, basil)
- Mushrooms
- Meat (jerky)
- Fish (salted and dried)
- Grains or soaked nuts for long-term storage

Drying prolongs the shelf life of food, concentrates its flavor, and makes emergency supplies, soups, stews, and snacks portable and always useable.

Each method—fermentation, drying, and canning—has unique advantages and a place in an off-grid food economy. Together, they allow you to preserve nearly any crop, such as vegetables from your garden, wild fruits, fish, and meat. You will need fewer freezers, make fewer trips to the grocery store, and enjoy the satisfaction of knowing that your shelves are stocked with food that you have grown, gathered, and stored yourself, ready to last through every season if you become proficient in these areas.

DIY Project 8:
Build a Survival Snare Trap

The skill to trap small game can be the difference between putting food on the table and going hungry when you're off the grid or in a survival situation. One of the earliest and most efficient methods for capturing rabbits, squirrels, and other small animals is to use a basic snare trap, which you can make with a few items you probably already own.

This do-it-yourself project demonstrates how to construct a simple snare trap that is simple to set, quick to deploy, and, with careful handling, reusable.

Tools & Materials:

- Thin, flexible wire (22 to 24 gauge works well) or strong paracord
- Knife or multi-tool
- A sturdy stick or small branch for the anchor stake
- Additional sticks to create a guide funnel (optional but helpful)

Step 1 — Find the Right Location

Start by scouting an area where a small game naturally travels. Look for well-worn animal trails, narrow funnels through the brush, or spots near food or water sources. Rabbits and squirrels follow the same paths, making these ideal places to set a snare.

The more natural the location, the better your chances. Setting the snare where animals are already moving reduces the need for bait.

Step 2 — Cut and Prepare Your Wire or Cord

Cut a wire or paracord between two and three feet long, depending on the game size you're aiming for. The snare will lock at this point, so form one end into a tiny loop and twist it tightly to secure it.

This is your snare; thread through the other end to make a wider loop that slips freely but tightens when pressure is applied.

The loop's diameter should be between 4 and 6 inches for rabbits or other animals of a comparable size. Ensure your snare is at head height for the species you aim for, usually 4 to 6 inches from the ground for rabbits.

Step 3 — Anchor the Snare

Sharpen one end of a strong stick or branch with your knife. This is your anchor point; drive it firmly into the ground next to the trail. Securely wrap the free end of your wire around the stake or fasten it with a paracord.

The snare must be sufficiently sturdy to prevent the animal from dragging it away after it has been caught.

Step 4 — Set the Snare in Position

Place the snare loop exactly where the animal will naturally walk—in the middle of the trail. If necessary, guide the animal toward the loop with little sticks on either side to improve your chances of a clean catch.

Verify that the loop is hanging freely and open, not touching the ground or catching on anything. Pulling the snare should cause it to tighten smoothly.

Step 5 — Check Your Trap Regularly

It is best to inspect a snare frequently. If you wait too long, your catch may spoil or be lost to other predators. If at all feasible, check traps once or twice a day.

To increase your chances, reset the trap as soon as you make a catch or if you see that it has been disturbed.

Chapter 9: Livestock Raising & Beekeeping

Keeping bees and raising livestock are two of the most practical and fulfilling ways to build a self-sufficient, off-grid homestead. Meat, milk, eggs, honey, wax, natural fertilizer, and even materials like leather and wool are produced by these living systems in addition to sustenance. More significantly, they establish a cycle of prosperity that strengthens your food security every year.

Off-grid animal husbandry lets you intimately connect to the land's rhythms. In contrast to corporate farming, it makes you consider breeding, water, shelter, and feed. Every humming swarm of bees, every fresh gallon of goat milk, and every chicken egg are the product of meticulous management and nature's cooperative efforts.

The fundamentals of choosing, rearing, and maintaining common livestock, such as pigs, goats, rabbits, and chickens, which provide food, compost, and even additional revenue, will be covered in this chapter. We'll also explore beekeeping, one of the most attractive and useful additions to any off-grid system. In addition to producing honey, bees pollinate your garden and maintain the health of your entire ecosystem.

Livestock and bees bring your off-grid lifestyle from survival to sustainability by providing a consistent, renewable supply of food directly from your property, regardless of how big or little you start.

Raising Chickens for Food Security

For good reason, chickens are frequently the first animals added to an off-grid homestead. They produce food rapidly, are adaptable, and require little upkeep. You can quickly get fresh eggs all year long with a small flock of hens, a great source of protein, good fats, and vital elements. In addition to producing eggs, chickens can be raised for their meat, providing a sustainable food supply with a comparatively quick turnaround time, mainly if you produce breeds with several uses.

They can live happily in a sturdy cop with a safe run because they don't require a lot of room. When working on the land, letting your birds roam freely reduces pest populations and lowers feed expenses by allowing your flock to forage naturally. Furthermore, chicken manure is a useful byproduct for composting or fertilizing your garden because it is high in nitrogen.

The speed at which chickens multiply is one of the best aspects of rearing them. Add meat birds as needed, or renew your flock by hatching new chicks every season with a broody hen or a tiny incubator. They are essential to food security since they are constantly active and creating something beneficial.

Raising Rabbits for Food Security

A rabbit is one of the most productive meat animals you can raise off the grid. They are peaceful, manageable, and prolific breeders, frequently giving birth to huge litters multiple times yearly. The fact that rabbits need less room and feed than larger livestock sets them apart. With little space requirements, a few well-run cages or colony arrangements can yield a consistent supply of lean, healthful meat.

Rabbit meat is excellent for roasting, stewing, and preserving because it is low in fat and protein and has a mild flavor. Due to their rapid growth, most meat breeds are ready for processing in 8 to 12 weeks. Because of this quick turnaround, you can modify your production to suit your requirements and the resources available.

The fact that rabbits flourish on hay, garden waste, and forage is an additional benefit. This enables you to feed them at a reasonable cost, mainly if you cultivate high-protein greens like alfalfa or clover. Their dung is a significant resource, much like chickens. Because rabbit droppings are "cold," they won't burn plants and can be applied straight to the garden, generating use from every part of the animal.

Raising Goats for Food Security

If you have a little extra area, goats are a great option for adding dairy and meat to your off-grid operation. Goats are resilient, inquisitive, and versatile animals that are natural browsers, meaning they will consume brush, weeds, and scrub that other animals would overlook. Because of this, they are excellent at clearing ground and turning rough feed into meat, milk, and even fiber.

One of the most adaptable products you can make is goat milk. It's better for creating cheese, yogurt, butter, and even soap, and it's easier to stomach than cow's milk. Your farm can become a miniature dairy with a healthy dairy goat, yielding up to a gallon of milk daily.

Boer goats are one example of a meat breed that grows quickly and offers a rich, tasty source of protein. A few homesteads raise goats for both milk and meat, using the same herd. Even if you don't want to prepare your own goats for meat, you can still maintain your system flexible and productive by trading or selling any extra offspring.

Because they are infamous escape artists, goats need sturdy fencing and simple shelter to keep out the weather. Goats are immensely fulfilling animals to raise if their needs are satisfied; they provide milk, meat, dung, and even company.

DIY Beehive Construction

Beekeeping is one of the most profitable additions to any off-grid farm. A healthy hive produces propolis, fresh honey, beeswax, and, most importantly, pollination for your garden and fruit trees. Making your own beehive allows you to customize it to your region and environment while saving money.

The top-bar hive is one of the most straightforward designs for do-it-yourself novices; it is simple to construct, check, and maintain without costly equipment. It's an excellent place to start with low-maintenance and ecological beekeeping.

Tools & Materials

- Untreated wood planks (cedar, pine, or fir work well)
- Screws or nails
- Saw, hammer, or drill
- Measuring tape and level
- Metal sheets or shingles for the roof
- Wood glue (optional)
- Hinges (if adding a hinged lid)

Step 1 — Build the Hive Body

A long, horizontal box, typically 3 to 4 feet long, 18 to 24 inches broad at the top, and 12 to 15 inches deep, should be the first thing you make. Cut the side panels into a trapezoid form by sloping them inward toward the bottom. This keeps the hive accessible for inspection and enables bees to spontaneously construct combs.

The ends must be sturdy enough to hold up the structure and flat. Make sure the box is level and robust before assembling it with screws or nails.

Step 2 — Add the Top Bars

The "top bars" are where your bees will attach their comb. Cut several wooden slats, about 1¼ to 1½ inches wide, long enough to rest across the width of the hive body. Space them evenly so they create an entire roof over the hive.

Each bar can contain a basic guide, such as a groove or a beeswax strip, to promote straight comb formation. Because these bars are detachable, honey collection and inspections are easier and cause less disturbance to the bees.

Step 3 — Build a Protective Roof

Protect your hive from rain and sun by building a simple pitched or flat roof that rests over the top bars. Use metal sheeting, shingles, or thick wood for weather resistance. If desired, attach the roof with hinges for easy access.

Make sure the roof extends past the hive walls to protect from rain. Good ventilation is important, so allow a small gap for airflow without letting in predators.

Step 4 — Create an Entrance

To allow the bees to enter and exit the hive, drill a tiny entrance hole (about 1 inch in diameter) on one end or along the bottom. As long as it's simple for the bees to defend, a single entrance or a few tiny holes are acceptable to some beekeepers.

Step 5 — Finish and Set Up

To preserve the bees' fragile wings, sand sharp edges inside the hive. Keep the inside natural and avoid painting or treating it. If necessary, untreated paint or natural oils might weatherproof the outside.

Place the hive on a stand about 18 inches off the ground to protect it from moisture, pests, and predators. Pick a sunny location facing southeast to capture the early light, offering wind shelter and afternoon shade.

DIY Project 9:
Build a Simple Beehive (Langstroth Design)

With good reason, the Langstroth beehive is among the most widely used hive designs among beekeepers. Its modular, stackable boxes give the bees a well-organized habitat while simplifying colony management, honey harvesting, and hive inspections.

Making your own Langstroth hive allows you to personalize it and save money. You can build a long-lasting hive with simple tools, perseverance, and high-quality wood.

Tools & Materials

- Wooden planks (pine or cedar works well)
- Nails or wood screws
- Hammer or drill
- Mesh wire (for the screened bottom board)
- Saw
- Measuring tape and level
- Wood glue (optional)
- Hinges (if adding a hinged lid)

Step 1 — Understand the Hive Components

Before building, it helps to know the basic parts of a Langstroth hive:

- **Bottom Board:** The hive's foundation — solid or screened for ventilation.
- **Deep Boxes (Brood Chambers):** Where the queen lays eggs, and the colony lives.
- **Medium or Shallow Boxes (Honey Supers):** Where bees store honey for harvest.
- **Frames:** Removable wooden frames where bees build comb.
- **Inner Cover and Outer Lid:** Protects the hive from weather and allows ventilation.

Aim for one or two deep boxes and one honey super to start for a simple build.

Step 2 — Build the Bottom Board

Cut a solid wood base large enough to fit your hive body — roughly 22" x 16". Attach three-sided wooden rails around the edges to create a landing board and prevent sliding.

If you want a screened bottom board for airflow, cut a center rectangle out of the base and staple mesh wire over the opening. This helps control mites and keeps the hive ventilated.

Step 3 — Construct the Deep Boxes (Brood Chambers)

Cut four wooden panels for each box:

- Two **22" side panels**
- Two **16" front/back panels**
- Depth: about **9½"**

Assemble the box into a rectangle, securing it with nails or screws. Make sure the corners are square, and the frame is solid. Repeat if building a second brood box.

Add a thin ledge or groove inside each box to hold the frames — these support the removable frames where bees build comb.

Step 4 — Build the Honey Super

The honey super is similar to the brood box but shallower — around **6½" deep.**

Repeat the steps above — cut, assemble, and install the frame ledge.

Step 5 — Assemble Frames

If building your own frames, create rectangular frames slightly smaller than the interior box dimensions. Install thin starter strips of wax or foundation mesh inside each frame to guide the bees as they build comb.

Pre-made frames can also be purchased to save time.

Step 6 — Build the Inner Cover and Outer Lid

For the inner cover, cut a flat panel of wood to fit snugly on top of the upper box. Drill a small hole for ventilation and bee access if you plan to feed them.

The outer lid should be slightly larger than the hive body, sloped or covered with metal sheeting to shed rain. Attach hinges if you want easy access, or rest it on top.

Step 7 — Stack and Secure the Hive

Once all parts are built, stack the components:

- Bottom board
- Deep box (or two)
- Honey super
- Inner cover
- Outer lid

Make sure everything fits snugly, but allow easy removal of each layer for inspection. Use straps or weights to keep the hive secure against strong winds.

Chapter 10: Advanced Gardening & Permaculture

Learning how to make your land work smarter, not harder, is the next step after mastering the fundamentals of gardening. To minimize waste and maximize harvests, advanced gardening techniques and permaculture concepts aim to create a self-sustaining system in which everything you plant, construct, and care for feeds into the following season.

Permaculture is a way of thinking, not merely a fancy phrase for gardening. It entails planning your homestead and garden to resemble the natural systems seen in nature. Animals, plants, soil, water, and even buildings all function in cycles, with one component supporting the others. This eventually results in a very productive, low-maintenance system that continues to pay you year after year.

This chapter explores companion planting, soil building, food forests, and water management techniques to help your garden flourish with minimal effort. You'll discover how to increase your food production, improve your soil, draw in helpful insects, and transform your homestead into a robust ecosystem that will provide for your family and rebuild the land you rely on.

At this point, gardening transcends survival and becomes a way of life.

Raised Bed Gardening (DIY)

One of the best methods for producing healthy, fruitful plants is raised bed gardening, mainly if your native soil is rocky, unfavorable, or challenging. You may enhance drainage, take control of the quality of your soil, and ease the strain of planting and harvesting by making your own raised beds.

The finest aspect? Once installed, raised beds will last many years and are easy to construct using simple tools and supplies.

Benefits of Raised Beds

- Better drainage — prevents waterlogging
- Warmer soil — extends your growing season
- Easier to manage soil fertility and weeds
- Ideal for areas with poor or compacted ground
- Allows dense planting — maximizing small spaces

Tools & Materials

- Untreated wooden planks (cedar, redwood, or pine work best)
- Screws or nails
- Drill or hammer
- Level, measuring tape, and saw
- Cardboard or landscape fabric (optional for weed suppression)
- Compost and rich garden soil mix

Step 1 — Choose the Right Location

Pick a sunny, flat area with good access to water. Most vegetables and herbs need at least 6–8 hours of sunlight daily, so choose your spot carefully.

Step 2 — Plan the Bed Size

Standard raised beds are **4 feet wide** — wide enough to plant generously but narrow enough to reach the center from either side without stepping on the soil. Length can vary — 8 to 12 feet long is common — and aim for a height of **12 to 24 inches** for deep roots and easy working height.

Step 3 — Cut and Assemble the Frame

Cut your wood to size. Assemble the rectangular frame by screwing or nailing the corners tightly. Reinforce longer beds with stakes or center supports to prevent bowing over time.

If you want extra strength or are working on uneven ground, use metal corner brackets or rebar for support.

Step 4 — Prepare the Ground

Clear any grass or weeds from the area. Optionally, lay down a layer of cardboard or landscape fabric to suppress future weeds and grass — this breaks down over time while protecting your plants' root zone.

Step 5 — Fill with Soil

Fill your raised bed with a rich mix of compost, topsoil, and organic matter. Aim for a fluffy, well-draining mix that holds moisture but doesn't stay soggy. If you have livestock, add aged manure to boost nutrients.

A good mix: **50% topsoil, 30% compost, 20% organic material (leaves, straw, aged manure).**

Step 6 — Plant and Maintain

Once filled, your raised bed is ready to plant. Densely plant your crops to shade out weeds and make the most of every square foot. Rotate crops yearly to keep the soil healthy.

Top off with compost each season and mulch to retain moisture and reduce weeds. Raised beds dry out faster than ground-level gardens, so check soil moisture regularly.

Pest Control & Organic Farming Techniques

One of the biggest challenges in off-grid or sustainable gardening is keeping crops healthy without reaching for chemical sprays. The good news? Nature already provides plenty of tools to help you manage pests and improve your soil — if you know how to use them.

Organic farming is not just about avoiding chemicals. It's about working with the land, building healthy soil, and creating a balanced environment where pests and diseases don't take over. The healthier your garden ecosystem, the less you'll need to fight off problems.

Start with Healthy Soil

The foundation of any organic garden is strong, living soil. Rich, well-fed soil grows stronger, and healthy plants naturally resist pests and diseases better. Adding compost, aged manure, and organic matter improves soil structure, feeds beneficial microbes, and gives your crops the strength they need to survive tough conditions.

Cover crops like clover, rye, or peas are also great tools. They protect your soil, prevent weeds, and pull nutrients deep into the ground — all while building fertility for the next planting season.

Attract Beneficial Insects

Not all bugs are bad. In fact, many insects are your best allies in pest control. Ladybugs, lacewings, praying mantises, and parasitic wasps all help keep harmful pests like aphids, caterpillars, and beetles in check.

Planting flowers and herbs like dill, fennel, yarrow, marigolds, and sunflowers near your garden draws these beneficial insects. The more natural predators you attract, the less you worry about infestations.

Companion Planting

Some plants naturally protect others when grown side by side. Companion planting is a simple, chemical-free way to deter pests while boosting your harvest.

For example:

- **Marigolds** repel nematodes and beetles.
- **Basil** protects tomatoes from flies and hornworms.
- **Garlic and onions** help drive off aphids and other soft-bodied insects.
- **Nasturtiums** are a trap crop, drawing pests away from your main vegetables.

Rotate crops yearly to prevent soil-borne diseases and pests from building up in one area.

Physical Barriers and Traps

Sometimes, the simplest solutions work best. Floating row covers, netting, and insect screens protect young plants from pests while letting in sunlight and rain. Mulching with straw or wood chips also helps block certain crawling pests while keeping soil moisture steady.

For slugs, try beer traps or hand-picking early in the morning. For larger invaders like rabbits or deer, fencing is often the best defense — tall enough to stop jumpers and buried deep enough to block diggers.

Organic Sprays and Remedies

When pests do show up, there are plenty of homemade or organic solutions:

- **Neem oil** is great against soft-bodied insects like aphids and mites.
- **Soap sprays** — a simple mix of water and natural dish soap suffocates pests on contact.
- **Diatomaceous earth** — harmless to people but deadly to bugs with exoskeletons.
- **Garlic and chili pepper sprays** — natural repellents that drive pests away without harming plants.

Use sprays early in the morning or late to protect beneficial insects.

Aquaponics & Hydroponics for Year-Round Food Production

When you're off the grid or aiming for total food security, growing fresh produce year-round can be a challenge — especially in harsh climates or during winter months. That's where **aquaponics** and **hydroponics** step in, offering innovative ways to grow food with fewer resources, no soil, and better control over your environment.

Both systems create closed-loop growing environments that maximize space, conserve water, and produce consistent harvests regardless of season.

Hydroponics — Growing Plants Without Soil

Hydroponics is the practice of growing plants directly in nutrient-rich water instead of soil. Roots are suspended in water or supported by an inert medium like clay pellets, perlite, or coconut coir. Nutrients are carefully added to the water, giving plants exactly what they need to grow fast and strong.

The benefits are clear:

- Faster growth compared to soil gardening
- No weeding, tilling, or soil pests to manage
- Precise control over nutrients and water use
- Perfect for indoor or greenhouse setups
- Allows vertical farming to maximize space

With hydroponics, you can grow leafy greens, herbs, tomatoes, peppers, and even strawberries — all year long, even in limited space.

The biggest challenge is keeping the system balanced. You'll need a basic understanding of pH levels, water temperature, and nutrient solutions — but once dialed in, hydroponics delivers consistent, high-yield crops that thrive indoors or out.

Aquaponics — Combining Fish and Plants in a Self-Sustaining System

Aquaponics takes hydroponics a step further by adding fish to the system. It creates a natural cycle where fish waste provides nutrients for the plants, and in turn, the plants filter and clean the water for the fish.

Here's how it works:

- Fish are raised in a tank — tilapia, catfish, trout, or even ornamental koi.
- Their waste, rich in ammonia, is pumped into grow beds where bacteria convert it into plant-friendly nutrients.
- Plants absorb those nutrients, cleaning the water before it cycles to the fish tank.

Aquaponics produces two food sources — fresh vegetables and fish — making it one of the most efficient off-grid food systems. It uses about **90% less water** than traditional gardening and produces protein and greens side by side.

Common aquaponic crops include:

- Lettuce, spinach, and kale
- Basil, mint, and other herbs
- Tomatoes, cucumbers, and peppers
- Edible flowers and some fruits

Meanwhile, fish like tilapia grow quickly and provide a renewable protein source.

Why These Systems Shine Off-Grid

Both hydroponics and aquaponics can run year-round with the help of solar power, small pumps, and grow lights if needed. They're perfect for greenhouses, small homesteads, or anywhere space and water are limited. Because they use controlled environments, pests and diseases are easier to manage — often eliminating the need for pesticides.

Aquaponics, in particular, creates a closed-loop system where waste becomes a resource — a true example of permaculture principles.

DIY Project 10: Build a Raised Bed Garden

A raised bed garden is one of the easiest ways to boost your growing space, especially if your soil is poor, rocky, or hard to work with. It gives you control over your soil quality, improves drainage, and makes planting, weeding, and harvesting much easier.

With a few basic tools and materials, you can build a raised garden bed that will serve you for years — helping you grow everything from vegetables and herbs to flowers and fruits.

Tools & Materials

- **Wood planks** (untreated cedar or pine works well — durable and safe for growing food)
- **Nails or wood screws**
- **Hammer or drill**
- **Measuring tape and saw**
- **Level (optional, but helpful)**
- **Soil mix** — rich garden soil blended with compost

Step 1 — Choose a Location

Pick a sunny, level spot for your garden bed — most vegetables need at least 6 to 8 hours of full sunlight daily. Ensure the area is accessible and close to a water source.

Step 2 — Measure and Cut the Wood

Decide on your bed's size. Standard and manageable size is **4 feet wide by 8 feet long and 12 to 18 inches tall** — wide enough to plant plenty but narrow enough to reach the center from either side without stepping into the bed.

Measure and cut your wood planks accordingly:

- **2 planks** for the long sides
- **2 planks** for the short ends

Step 3 — Assemble the Frame

Lay the boards out on flat ground to form a rectangle. Nail or screw the corners together tightly — adding a corner stake or brace inside each corner for extra strength if needed.

Use a level to make sure your frame is sitting flat and square. This will help with water drainage and prevent warping over time.

Step 4 — Prepare the Ground

Clear any grass, weeds, or large rocks inside the frame area. If you're worried about weeds creeping in, lay down a layer of **cardboard or landscape fabric** at the bottom of the bed before adding soil. This will break down over time while keeping weeds at bay.

Step 5 — Fill with Soil

Fill your raised bed with a rich blend of **topsoil, compost, and organic matter.** A good ratio is:

- 50% topsoil
- 30% compost
- 20% **organic material** like aged manure, leaf mold, or peat

Level the soil and lightly tamp it down. The soil should be loose but not too fluffy — ready for planting.

Step 6 — Plant and Maintain

Your raised bed is now ready! Plant your crops densely but carefully — raised beds allow for tighter planting because of improved soil quality and drainage.

Add a layer of mulch to help retain moisture and reduce weeds. Check moisture levels often, as raised beds can dry out faster than in-ground gardens.

Top off the soil with compost each season to keep the nutrients flowing and the garden productive.

Advanced Gardening & Permaculture

End-of-Chapter Checklist: Securing Your Food Supply

- ☑ **Gardening System in Place** — Have you established a reliable garden, raised beds, or permaculture setup that produces seasonal crops?

- ☑ **Livestock or Small Animals Integrated** — Are chickens, rabbits, goats, or other small livestock part of your food system that provide meat, milk, eggs, or fertilizer?

- ☑ **Foraging, Hunting, and Trapping Skills Practiced** — Have you identified local wild edibles, learned basic trapping, and developed responsible hunting skills for added protein and seasonal harvests?

- ☑ **Food Preservation Methods Ready** — Can you ferment can, dehydrate, or smoke food to stock your pantry through lean seasons?

- ☑ **Beekeeping or Pollination Plan Active** — Have you created beehives or a pollinator-friendly garden to support healthy crops and harvest honey?

- ☑ **Aquaponics or Hydroponics Considered** — If space or climate is challenging, are you exploring water-based systems for year-round food production?

- ☑ **Organic Pest Control and Soil Health Management** — Are you using natural pest control, building soil fertility, and rotating crops to protect your long-term harvests?

- ☑ **Seed Saving or Heirloom Crops Started** — Are you collecting seeds or growing heirloom varieties to ensure your garden is self-sustaining year after year?

- ☑ **Food Storage System Set Up** — Do you have shelves, root cellars, or storage spaces ready for dried goods, canned foods, and preserved harvests?

Section 4:
HEALTH, HYGIENE & EMERGENCY PREPAREDNESS

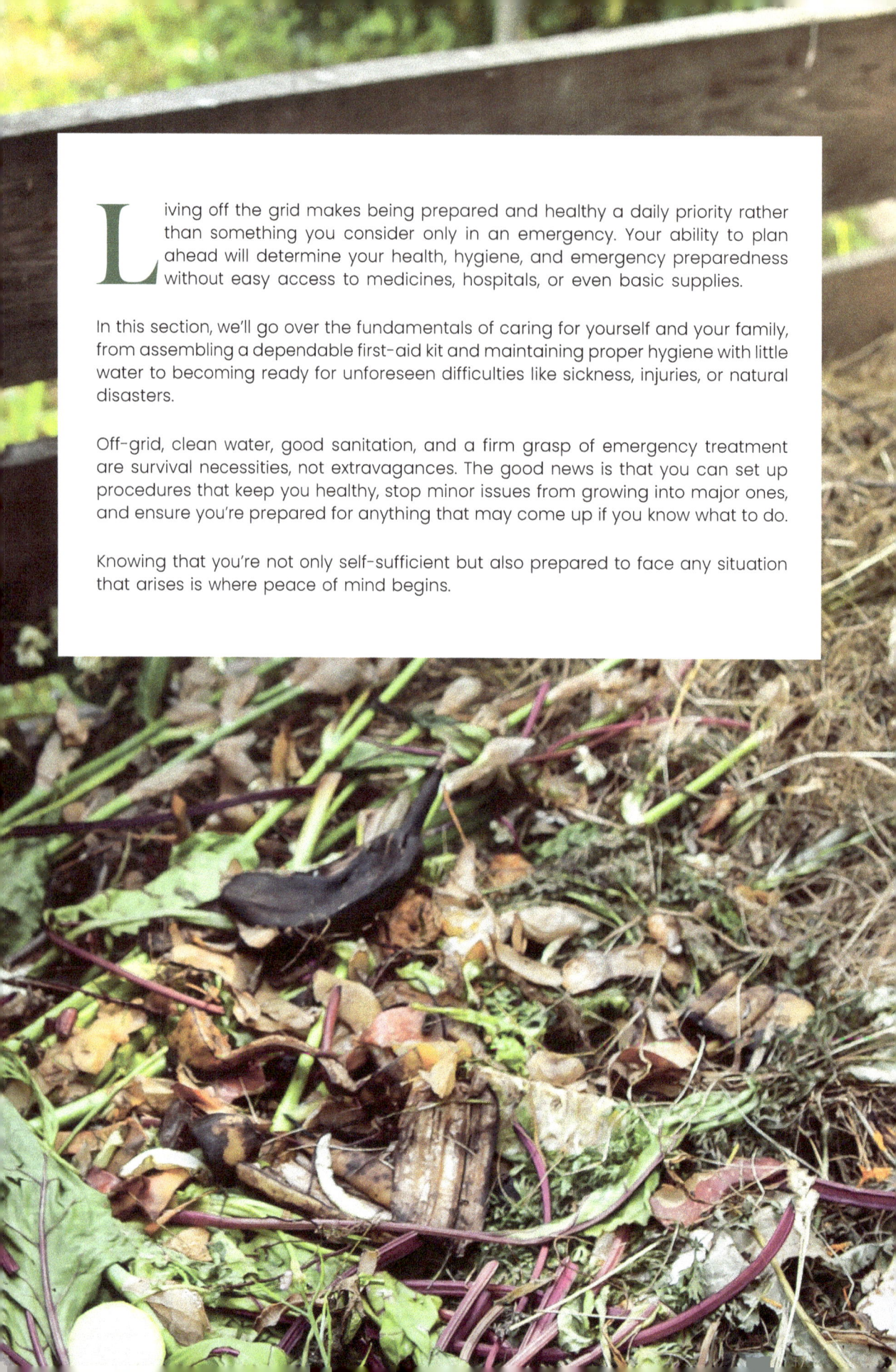

Living off the grid makes being prepared and healthy a daily priority rather than something you consider only in an emergency. Your ability to plan ahead will determine your health, hygiene, and emergency preparedness without easy access to medicines, hospitals, or even basic supplies.

In this section, we'll go over the fundamentals of caring for yourself and your family, from assembling a dependable first-aid kit and maintaining proper hygiene with little water to becoming ready for unforeseen difficulties like sickness, injuries, or natural disasters.

Off-grid, clean water, good sanitation, and a firm grasp of emergency treatment are survival necessities, not extravagances. The good news is that you can set up procedures that keep you healthy, stop minor issues from growing into major ones, and ensure you're prepared for anything that may come up if you know what to do.

Knowing that you're not only self-sufficient but also prepared to face any situation that arises is where peace of mind begins.

Chapter 11: Off-Grid Hygiene & Waste Management

Hygiene and trash management are two aspects of off-grid living that are frequently disregarded until necessary. Maintaining the cleanliness of your surroundings, family, and yourself is essential for survival and health, not only for comfort. Waste rapidly becomes a hazard without the proper procedures, drawing bugs, poisoning water, and spreading illness.

Building a Composting Toilet

Managing human waste appropriately is imperative when living off the grid, and digging a pit isn't a sustainable answer. Composting toilets can help with that. A composting toilet converts waste into useful compost while maintaining a clean and hygienic environment. It is easy to install and maintain and is environmentally beneficial.

How It Works

With the help of carbon-containing materials like sawdust, a composting toilet breaks down human waste into safe, usable compost over time by separating solid and liquid waste. The process doesn't require plumbing or water or produce any unpleasant odors when done correctly.

Tools & Materials

- Plywood or wooden boards (for the frame and box)
- Toilet seat (store-bought or homemade)
- 5-gallon buckets or containers (for waste collection)
- Sawdust, straw, or dry leaves (carbon material)
- Screws, nails, drill, saw
- **Optional:** urine diverter for separating liquids

Step 1 — Build the Toilet Frame and Seat

First, construct a basic wooden box tall enough to accommodate a 5-gallon bucket. Make a cut on the top that is large enough to accommodate your toilet seat. You can make your own seat out of wood or purchase a stock for a more rustic style. Ensure the box is strong enough to comfortably hold a person's weight.

Step 2 — Set Up the Waste Bucket

The 5-gallon bucket should be placed inside the box below the seat opening. Some designs incorporate a urine diverter that directs liquid waste to a different container or drain to lessen odor and facilitate solids composting. This is helpful but optional.

Beside the toilet, place a container of dry carbon material, such as sawdust, straw, shredded leaves, or even dry peat moss.

Step 3 — Start Using the System

Cover solid trash with a liberal scoop of your carbon material after each usage. To properly compost, this absorbs moisture, stops odors, and balances the nitrogen in human waste.

If you maintain equilibrium, nature will handle the breakdown without chemicals or store-bought items.

Step 4 — Empty and Manage Compost

Once the bucket is full, seal it and move it to the composting site, a designated pile or bin away from your garden and water sources. Allow it to settle and break down for at least six months to a year. Over time, the trash will decompose into wholesome, safe compost.

Food crops should never be directly treated with humanure compost. Use it on trees, bushes, and non-edible plants.

Natural Soap & Toothpaste Recipes

Making your own soap and toothpaste when off the grid is about being independent, not only about avoiding harsh chemicals. You may make natural products that keep you clean and healthy while being kind to the earth and your body with a few simple components. Additionally, it saves you trips to the market and is an easy and enjoyable process.

Natural Soap Recipe — Basic Cold-Process Soap

Homemade soap is surprisingly easy once you learn the basics. This recipe uses natural fats and oils to create a moisturizing, chemical-free soap for everyday use.

Ingredients:

- 16 oz. olive oil (or a mix of olive, coconut, and lard/tallow)
- 6 oz. coconut oil (for lather and hardness)
- 2 oz. castor oil (optional, boosts bubbles)
- 6 oz. distilled water
- 2.3 oz. lye (100% sodium hydroxide)
- Essential oils for scent (lavender, tea tree, or peppermint work well)
- Dried herbs, oats, or clays (optional, for texture or skin benefits)

Instructions:

1. **Safety first** — Wear gloves and eye protection. Always add lye to water (never the other way around) and mix outdoors or in a well-ventilated area.
2. Slowly mix the lye into the water. Stir until dissolved and set aside to cool.
3. In a separate pot, gently heat your oils until thoroughly melted, then cool to about 100°F.
4. Once the lye water and oils are around the same temperature (90-100°F), slowly pour the lye water into the oils.
5. Use a stick blender or stir by hand until the mixture thickens (called "trace").
6. Add essential oils or additives, mix well, then pour into molds.
7. Cover and let it set for 24–48 hours. Once solid, remove from molds and cure the bars in a cool, dry place for 4–6 weeks.

The ultimate product is an all-natural, skin-friendly soap that you can use for laundry, dishwashing, and everyday hygiene.

Natural Toothpaste Recipe — Simple and Effective

Commercial toothpaste often contains artificial sweeteners, fluoride, and preservatives. Making your own natural version is quick, inexpensive, and keeps your teeth clean without unnecessary chemicals.

Ingredients:

- 4 tbsp. baking soda (gentle abrasive)
- 2 tbsp. coconut oil (antibacterial and adds smooth texture)
- 1 tbsp. bentonite clay (optional for added minerals and detox)
- 10–15 drops peppermint or spearmint essential oil (for freshness)
- 1 tsp. Fine sea salt (optional for minerals and mild abrasiveness)

Instructions:

1. Mix all ingredients in a small bowl until it forms a smooth paste.
2. Adjust coconut oil or baking soda to get the consistency you like.
3. Store in a small jar with a lid. Scoop a small amount onto your toothbrush as needed.

This natural toothpaste gently cleans teeth, freshens breath, and supports oral health without synthetic ingredients. Coconut oil is antibacterial, while baking soda and salt help polish and remineralize.

DIY Project 11:
Construct a Composting Toilet

A composting toilet is among the most useful off-grid projects you will construct. It is easy to use, uses little materials, and transforms waste management into a sustainable, odorless system that doesn't require water, chemicals, or plumbing.

This DIY version uses the "bucket system" — affordable, easy to build, and fully functional for daily use.

Tools & Materials

- **5-gallon bucket** (or larger)
- **Wooden planks or plywood** (for building the frame and seat platform)
- **Toilet seat cover** (store-bought or homemade)
- **Saw, drill, screws, measuring tape**
- **Hinges** (optional if you want a liftable lid)
- **Carbon material:** sawdust, dry leaves, straw, or peat moss (for covering waste)

Step 1 — Build the Wooden Frame

Start by building a sturdy wooden box large enough to house your bucket but tall enough for comfortable sitting — usually around **16 to 18 inches high.**

Cut a hole in the top platform to fit the toilet seat or bucket rim snugly. This will hold your seat in place and support the user's weight.

Optional: Add side panels or a hinged lid if you want a cleaner look or easier access to the bucket.

Step 2 — Attach the Toilet Seat

Secure your toilet seat over the hole. If using a store-bought seat, simply screw it onto the wood. Cut a wooden seat with a smooth opening for a more rustic design.

Make sure everything is sturdy and well-attached — the last thing you want is wobbling or shifting when in use.

Step 3 — Insert the Bucket

Place your bucket directly underneath the seat opening. Make sure it sits firmly and can be easily removed when full.

Keep a container of your **carbon material** — sawdust, straw, or dried leaves — nearby. This is essential for covering waste, controlling moisture, and eliminating odors.

Step 4 — Using the Composting Toilet

After each use, add a generous scoop of sawdust or carbon material to cover the waste entirely. This keeps the system odor-free and balances composting by soaking up liquids and adding necessary carbon.

Empty the bucket into a designated composting area when complete. The system should be clean, dry, and low-maintenance if appropriately managed.

Chapter 12: Natural Remedies & First Aid

Living off the grid might limit your access to healthcare, so it's critical to understand how to take care of your family and yourself in the event of an illness or injury. You can directly manage anything from small wounds to major illnesses on your farm using natural treatments, herbal knowledge, and a well-stocked first aid box.

This chapter gives you the knowledge and skills to manage common health conditions organically and on your own by covering the fundamentals of herbal medicine, how to assemble an off-grid first aid kit, and a practical project to make your own herbal treatments.

Herbal Medicine Basics & Emergency Herbal Kits

When you're off the grid, nature becomes your pharmacy and your first line of defense. Herbal medicine has been used for centuries to treat wounds, illnesses, and everyday aches long before modern medicine existed. Knowing a few key plants and how to prepare them means you can handle most minor health issues right from your land.

Understanding Herbal Medicine Basics

At its core, herbal medicine is about working with plants containing natural compounds that support the body's healing processes. Many common herbs have antibacterial, anti-inflammatory, or immune-boosting properties — and learning how to use them safely gives you a considerable advantage when professional help isn't close by.

Natural Remedies & First Aid

Herbs can be used in several forms:

- **Infusions or Teas** — for soothing fevers, calming nerves, or aiding digestion
- **Tinctures** — concentrated herbal extracts preserved in alcohol for long-term use
- **Salves and Balms** — for cuts, scrapes, bruises, and skin irritations
- **Poultices** — fresh or dried herbs applied directly to the skin to draw out infection or reduce swelling
- **Syrups** — great for coughs and sore throats and boosting the immune system

Knowing how to prepare these is part of building true off-grid resilience. Even a small collection of herbs can replace dozens of store-bought medications.

Key Medicinal Herbs to Include

If you're just starting out, here are a few powerful herbs worth drying, growing, or keeping in your emergency herbal kit:

- **Yarrow** — Stops bleeding, reduces fever, helps heal wounds
- **Plantain (broadleaf or narrow-leaf)** — Soothes bites, stings, rashes, and pulls out toxins
- **Calendula** — Speeds healing of cuts, burns, and skin irritations
- **Comfrey (use externally)** — Helps heal bruises, sprains, and broken bones
- **Echinacea** — Immune booster, fights infections
- **Chamomile** — Calming, good for digestion and sleep
- **Elderberries** — Support immune health and help fight colds and flu
- **Peppermint** — Soothes upset stomachs and headaches
- **Lavender** — Calming, helps with burns, sleep, and anxiety

Building an Emergency Herbal Kit

An off-grid herbal kit should be compact but packed with versatile remedies ready for quick use. Here's what to include:

- **Dried Herbs** — Stored in airtight jars for teas, poultices, or infusions
- **Tinctures** — Alcohol-based herbal extracts that last for years
- **Healing Salves and Balms** — Calendula or plantain salve for wounds and skin
- **Essential Oils** — Lavender, tea tree, and peppermint for first aid and relaxation
- **Activated Charcoal** — For food poisoning, bites, or stings
- **Electrolyte Mix or Herbal Rehydration Formula** — For dehydration
- **Basic Tools** — Small scissors, tweezers, gauze, and bandages for applying herbs or poultices

Label everything clearly and pack your kit in a waterproof, portable bag so it's ready to grab in an emergency.

Essential Off-Grid First Aid Kit

There is no assurance that medical assistance will be available when you are off the grid. This implies that your first aid pack should have more than just the bare minimum; it should be designed with the idea that you will need to handle situations on your own until assistance arrives or things settle down.

Being ready for minor cuts as well as more serious injuries, including burns, deep wounds, sprains, and infections, is the aim. Your first aid kit might become your lifeline if you have the necessary equipment and a little expertise.

Basic Wound Care Supplies

- **Sterile gauze pads (various sizes)** — For covering wounds and absorbing blood
- **Rolled gauze and bandages** — For wrapping wounds, sprains, or securing splints
- **Medical tape** — To secure bandages or dressings
- **Adhesive bandages** — For minor cuts and scrapes
- **Antiseptic wipes or iodine solution** — To clean wounds and prevent infection
- **Tweezers** — For removing splinters, ticks, or debris
- **Scissors** — For cutting gauze, tape, or clothing if needed
- **Gloves (nitrile or latex)** — To protect yourself and keep wounds clean

Burn and Bite Treatment

- **Burn gel or aloe vera gel** — For minor burns
- **Honey (medical-grade or raw)** — Natural antibacterial for burns and wounds
- **Baking soda or clay powder** — Useful for insect bites, stings, or mild skin irritations
- **Benadryl or antihistamine cream** — For allergic reactions or itchy rashes
- **Activated charcoal** — For snake bites, food poisoning, or detox needs

Pain Relief and Inflammation

- **Ibuprofen, aspirin, or acetaminophen** — For pain, fever, and inflammation
- **Arnica gel or salve** — Helps with bruises, sore muscles, and swelling
- **Epsom salts** — For soaking sore muscles or drawing out infection from wounds

Serious Wound & Trauma Supplies

- **Suture kit or butterfly strips** — For closing deep wounds when stitches aren't available
- **SAM splint or stiff board** — For immobilizing broken bones or sprains
- **Israeli bandage or compression bandage** — For stopping serious bleeding
- **Tourniquet** — A last-resort tool for life-threatening bleeding
- **Eyewash or saline solution** — For flushing eyes or cleaning deep wounds

Medications & Extras

- **Electrolyte packets** — For dehydration, heat exhaustion, or illness
- **Anti-diarrheal medication** — Useful when clean water is scarce
- Antibiotic ointment or powder — For preventing infections in minor wounds
- **Herbal tinctures (optional)** — Yarrow for bleeding, echinacea for infection, chamomile for calming
- **Thermometer** — For monitoring fevers
- **Emergency blanket** — For shock or cold weather emergencies
- **Basic first aid manual or quick reference guide** — For guidance under pressure

DIY Project 12:
Make a Natural Herbal First Aid Kit

One of the most empowering off-grid projects you can do is make your own herbal first aid pack. When minor wounds, bites, or common illnesses occur, you will have natural salves, tinctures, and dried herbs instead of depending on store-bought medicines that are heavy in chemicals.

With a few jars, some common herbs, and beeswax, you can create a basic, sustainable kit that you can quickly restock from your land over time.

Tools & Materials

- Small **glass jars or tins** (for salves and dried herbs)
- **Dried medicinal herbs** (like calendula, plantain, yarrow, comfrey, chamomile, lavender)
- **Beeswax** (for salve-making)
- **Olive oil or coconut oil** (as a carrier oil)
- Cheesecloth or fine strainer
- Labels and a marker
- A sturdy container or bag to store everything

Step 1 — Choose Your Core Herbs

Pick herbs that cover a range of first-aid uses. Some of the most versatile options include:

- **Calendula:** For cuts, burns, and skin irritations
- **Plantain:** Pulls out toxins from stings or bites
- **Yarrow:** Stops bleeding and aids wound healing
- **Comfrey:** Supports healing of bruises, sprains, and strains (external use only)
- **Chamomile:** Calming, good for skin, and soothes upset stomachs
- **Lavender:** Antiseptic and calming, good for burns and stress relief

Keep the herbs dried in labeled jars or bags, ready for teas, poultices, or infusions.

Step 2 — Infuse Your Herbal Oil

For healing salves, you'll need herb-infused oil:

1. Place dried herbs into a clean jar.
2. Cover completely with olive oil or coconut oil.
3. Let it sit in a sunny window or warm spot for 2–3 weeks, shaking daily.
4. Strain the oil through cheesecloth into a clean jar.

This infused oil becomes the base for your salves.

Step 3 — Make a Healing Salve

Now, turn your infused oil into a ready-to-use salve:

1. Measure **1 part beeswax** to **4 parts herbal oil.**
2. Melt together slowly over low heat, stirring until smooth.
3. Pour into small jars or tins and let it cool until solid.
4. Label each jar with its intended use — like "Calendula Healing Salve" or "Plantain Bite & Sting Balm."

Step 4 — Add Herbal Tinctures or Powders (Optional)

If you have tinctures — like echinacea for immunity or yarrow for fever — add them to your kit. You can also store powdered herbs like slippery elm for digestive issues.

Step 5 — Assemble and Label

Pack everything into a waterproof bag or sturdy box.

- Healing salves
- Dried herbs
- Tinctures or oils
- Cheesecloth, dropper bottles, small spoon
- Labels and a quick herbal use guide

End-of-Chapter Checklist: Your Health & Emergency Preparedness Plan

- ☑ **Composting Toilet or Sanitation System Built** — Do you have a reliable, safe method for managing waste to protect your health and water sources?

- ☑ **Basic Hygiene Supplies Ready** — Have you prepared essential items like soap (store-bought or homemade), natural toothpaste, clean water storage, and sanitation tools?

- ☑ **Natural Remedies & Herbal Kit Assembled** — Is your herbal first aid kit stocked with dried herbs, salves, tinctures, and supplies for treating common injuries or illnesses?

- ☑ **Off-Grid First Aid Kit Fully Stocked** — Have you included medical essentials like gauze, antiseptics, suture tools, pain relievers, and trauma supplies in your kit?

- ☑ **Emergency Water Purification Methods Secured** — Are you prepared to clean and filter water in a health emergency with tools like filters, charcoal, or boiling setups?

- ☑ **Basic First Aid and Herbal Knowledge Practiced** — Have you learned or practiced first aid skills and simple herbal remedies so you're confident using them?

- ☑ **Emergency Plan Written or Reviewed** — Does everyone in your household know the plan for handling injuries, illnesses, or extreme weather events while off-grid?

- ☑ **Backup Power or Lighting Ready for Medical Needs** — Have you accounted for lighting or power needs if treating wounds or emergencies happens at night?

Section 5:
SECURITY & LONG-TERM RESILIENCE

Building an off-grid lifestyle involves more than just finding clean water, producing electricity, or growing food; it also entails safeguarding what you've put so much effort into creating. In actuality, there are risks associated with self-sufficiency. Securing your home, supplies, and mental health is just as crucial as keeping your pantry stocked, regardless of the threats you face—wild animals, opportunistic burglars, or living alone.

In this section, we'll go over doable strategies for fortifying your off-grid farmhouse, such as installing early warning systems, installing a smart fence that blends in with the surroundings, and strengthening windows and doors. Resilience, however, is more than simply physical. It's also mental. Living off the grid frequently entails dealing with extended periods of isolation, unforeseen difficulties, and the strain of being totally independent. Thus, developing mental toughness and locating (or establishing) a group of people with similar values can make the difference between thriving and merely getting by.

Furthermore, we will discuss realistic, doable initiatives like creating your own barter kit and setting up simple security alarms—small steps that make a big difference in your long-term resilience. Because being really off the grid means not only being ready for today but also for whatever the future may bring.

Chapter 13: Off-Grid Security & Home Defense

You become your own first line of defense when you live off the grid, against the dangers of nature but also against human attacks. You are responsible for safeguarding your house, belongings, and family without the protection of local law enforcement or contemporary security systems.

The fact is that your homestead may become a target in an off-grid environment, particularly in times of crisis or scarcity. Your property must be safe enough to provide you with peace of mind day and night, regardless of the threats it faces—from inquisitive trespassers to desperate people or even untamed animals.

The main subject of this chapter is the practical methods of fortifying your house and property without using sophisticated equipment. We'll talk about securing entrance points, erecting sturdy fences, hiding with vegetation, and utilizing natural allies like guard dogs and early warning systems. You'll have a firm grasp of how to build security layers that safeguard the most important things by the end, allowing you to sleep quickly, knowing your homestead is just as secure and independent as the rest of your off-grid existence.

Fortifying Doors & Windows

Regarding off-grid security, your doors and windows are the first and most obvious entry points. These aren't just access points for people — they're also the easiest targets for wildlife or weather damage if not properly reinforced. Fortifying them isn't about making your home look like a fortress but about smart, practical upgrades that keep you safer without drawing unnecessary attention.

Securing Doors

Start with your main doors. If you're using lightweight or hollow-core doors, replacing them with solid wood or steel-core doors is the first significant upgrade. A solid door instantly boosts your home's security and makes forced entry much harder.

The door frame is just as important. Reinforce it with heavy-duty strike plates and long screws that sink into the wall studs — not just the door frame itself. This small detail makes it much harder for someone to kick in your door.

Consider installing a sturdy deadbolt and a security bar or door brace, especially for nights or when you're away. A simple wooden bar or metal brace wedged against the door can stop an intruder, giving you precious time to react.

For outbuildings, sheds, or garages, don't overlook their doors. Those often get neglected but might be where you store fuel, tools, or other critical supplies. Apply the same upgrades — solid doors, heavy locks, and reinforced frames.

Reinforcing Windows

Windows provide easy access, light, ventilation, and views but are always weak. Strengthening your windows just means being wise about guarding them; sacrificing those advantages doesn't require sacrificing.

Adding security film, a clear coating that keeps glass together if broken, is one of the easiest additions. Although it doesn't prevent impact, it saves time by preventing the glass from breaking immediately.

Put in hinged panels or shutters made of solid wood that can be closed and secured from the inside. Closing your shutters during a storm or other high-risk event provides additional protection from flying debris and humans.

Another choice, particularly for isolated locations, is window bars or grates. Many are ornamental but still sturdy enough to prevent someone from climbing through, so they don't have to resemble prison bars if they're properly made.

Lastly, placing prickly shrubs like roses, blackberries, or hawthorn under windows offers a true barrier and a natural deterrent that complements your landscape.

DIY Perimeter Fencing & Concealment Landscaping

Your property is your lifeblood when you're living off the grid; it contains your house, food, water, and everything you've worked so hard to accumulate. Controlling the perimeter is the first step in protecting that area. In addition to defining your boundaries, a well-designed fence and thoughtful landscaping deter intruders, impede possible threats, and even make your farm fit in with the surrounding area.

Building DIY Perimeter Fencing

Fencing only needs to be sturdy, strategically located, and appropriate for your particular piece of land; it doesn't need to be ornate or costly. To begin, stroll your property line and look for weak points, such as open fields, low regions, or secret paths where someone (or something) could get through.

For a simple yet powerful off-grid fence, think about utilizing:

- **Wooden posts and rails** — sturdy, natural-looking, and easy to repair.
- **Welded wire or field fencing** — keeps out deer, stray dogs, or even people if tall enough.
- **Barbed wire or electric strands** — add a serious deterrent if you're in a high-risk area (check local regulations).

Because they bear the weight of the entire fence line, make sure you brace and sink your corner posts deeply during installation. Stretch wire is enough to maintain its shape if you add it, but not so much that it breaks under pressure.

Gates should be sturdy, locked, and sufficiently broad for equipment if necessary. Simple wooden gates that are strengthened with metal brackets are effective. For strength, use cross braces and sturdy hinges.

Concealment Landscaping — Blending In While Defending

Having visibility has two drawbacks. You don't want everyone to see you, but you want to see what's coming. Planting trees, shrubs, and other natural obstacles to cover your house and activities without drawing attention is known as concealment landscaping.

Thorny bushes, such as wild roses, hawthorn, or blackberries, are ideal since they are hard to push through, dense, and grow quickly. Plant an additional natural defense layer under windows or along fence lines if you want an additional natural defense layer.

All year long, thick trees or evergreens provide visual barriers. You can blend your property into the surrounding scenery and obscure roadways or neighbors' sightlines with a row of spruce, cedar, or pine.

In addition to shade, earth berms, or natural mounds can enhance drainage and provide wind protection. They blend with the landscape when topped with grasses or shrubs.

Think about establishing "choke points"— narrow walkways that direct foot traffic precisely where you want it, such as toward gates or open areas that are easy to watch. It becomes simpler to identify strangers or dangers the more you direct movement.

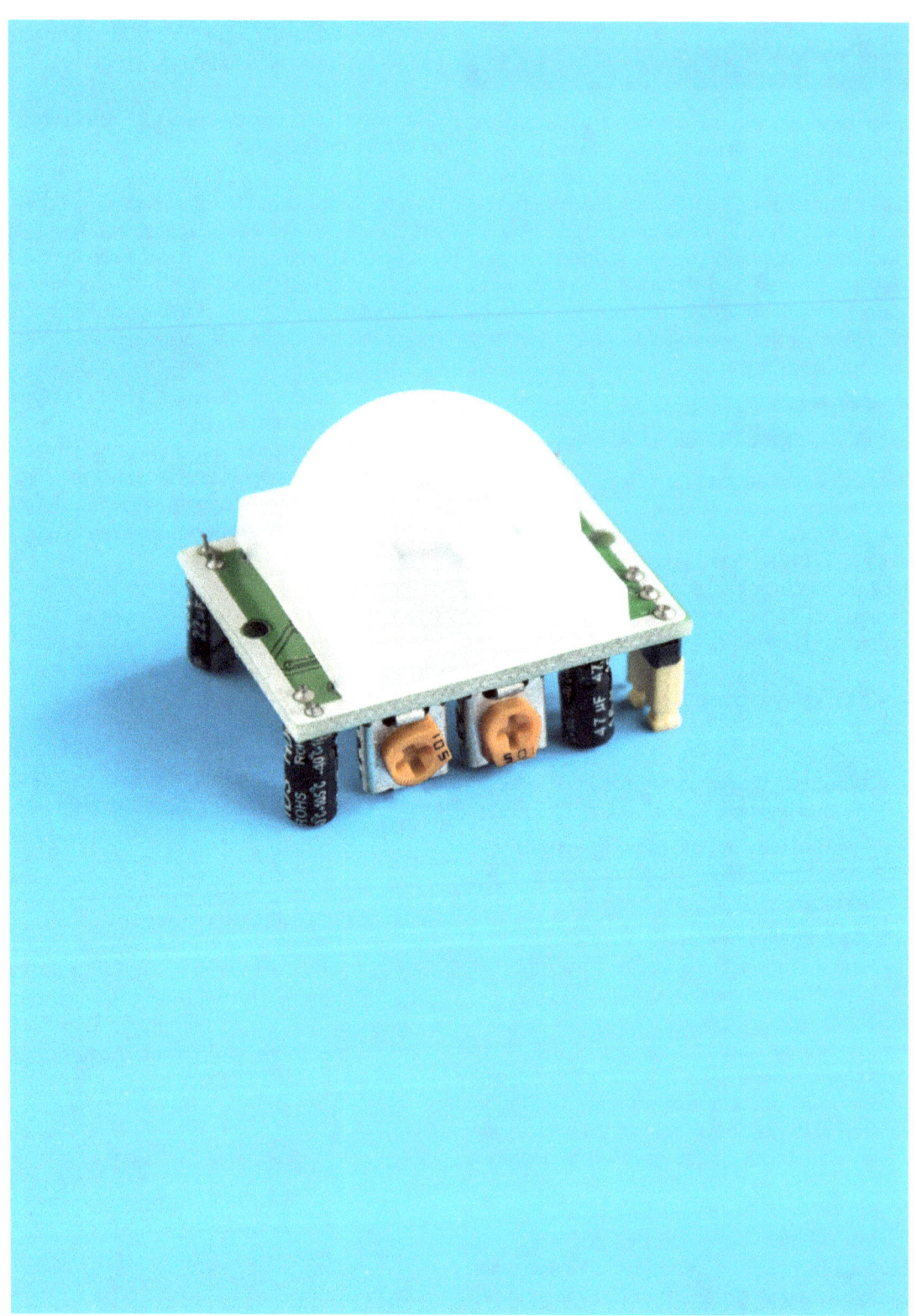

DIY Project 13:
Install a Simple Motion Sensor Alarm

Installing early warning systems that notify you before someone (or something) approaches too closely is essential to protecting your off-grid farm. One of the simplest and least expensive methods to increase security is with a basic motion sensor alarm.

Without using the internet or complex systems, you may set up an alarm that will notify you of movement in your house, garden, or animal areas with a few simple supplies.

Tools & Materials

- **Battery-powered motion sensor alarm kit** (available at most hardware or farm supply stores)
- **Screwdriver or drill**
- **Screws or mounting hardware**
- **Rechargeable batteries or solar-powered motion sensors** (optional but ideal for off-grid use)
- **Optional:** Weatherproof casing if your alarm isn't designed for outdoor use

Step 1 — Choose the Right Motion Sensor

For off-grid use, use a battery-operated model or one that charges via solar power. Some models emit a loud siren, while others send a chime or even a silent alert to a handheld receiver. Choose what works best for your setup — loud for perimeter defense or quiet if you just want a heads-up.

Ensure the sensor has a decent range — many basic models cover **20 to 30 feet** — and check the detection angle.

Step 2 — Select the Installation Spot

Walk your property and decide where you need coverage most:

- Entryways, driveways, or main paths leading to your home
- Near outbuildings, tool sheds, or fuel storage
- Garden or livestock areas vulnerable to animals or trespassers

Mount the sensor about 4 to 5 feet off the ground — high enough to avoid triggering from small animals but low enough to catch human movement. Aim the sensor toward the approach, not directly at sunrise or sunset, to avoid false alarms from light changes.

Step 3 — Install the Sensor

Secure the sensor using screws or the provided mounting hardware. Most units are simple to install and adjust.

If using rechargeable batteries, ensure the sensor is exposed to enough sunlight if it's solar-powered. Check that the battery compartment is sealed against moisture if placed outdoors for battery-only models.

Step 4 — Test the System

Walk through the sensor zone to test its range and sensitivity. Adjust the angle to avoid false triggers from blowing branches or small animals.

Confirm that the alarm — whether a sound, chime, or alert — is loud and clear enough to hear from wherever you plan to be.

Step 5 — Maintain Regularly

Check the batteries or solar charge regularly, especially before winter or storm seasons. Clear away debris or branches that might interfere with the sensor's view.

Chapter 14: Mental Resilience & Community Building

Living off the grid is as much a mental as a physical experience. It's easy to undervalue the emotional and psychological difficulties of living aside from contemporary society, even though most people concentrate on necessities like food, water, and shelter. If you're not ready for it, loneliness, extended periods of alone, unforeseen failures, and the weight of duty can be taxing.

Because of this, mental toughness is as essential to your long-term survival as any other item in your toolbox. It's about finding happiness and meaning in the daily grind of independent life, handling difficult situations, and maintaining composure when plans go awry.

However, being resilient does not entail working alone. Whether it's a network of homesteaders or a few neighbors down the street, establishing a strong, like-minded community builds a safety net that benefits all parties. In addition to shared resources, abilities, and information, the community also entails friendship, support during difficult times, and the simple awareness that you're not alone.

Because surviving off the grid is difficult, but thriving off the grid requires a village, we'll look at how to safeguard your mental health, deal with stress, and build enduring relationships with others who respect self-reliance, perseverance, and hard work in this chapter.

Coping with Isolation & Stress

One of the most complex parts of off-grid living — and something most people don't talk about — is the mental weight of isolation. The quiet that once felt peaceful can, over time, feel heavy. Days and weeks without much outside contact can lead to restlessness, anxiety, or even depression if you're not careful.

Stress is inevitable when you're responsible for every aspect of your survival — from growing your food to fixing what breaks. There are no repairmen on call, no quick trips to the store, and sometimes, no one around to lend a hand. When challenges stack up, the feeling of being genuinely alone can hit harder than expected.

Recognizing the Signs

The first step in coping is knowing the signs that stress or isolation are creeping in:

- Feeling irritable or overly anxious
- Losing interest in tasks you once enjoyed
- Sleeping too much or struggling to sleep at all
- Constant worry about things beyond your control
- A growing sense of loneliness, even in the quiet moments

These feelings are normal — they don't mean you're not cut out for this life. They mean you're human.

Healthy Ways to Cope

Instead of trying to power through or ignore the mental strain, it helps to build routines and habits that protect your mind just as you protect your body.

1. **Stick to Daily Routines:**

    Structure keeps your mind steady. Even something simple — morning coffee, evening chores, or a daily walk — creates a rhythm that grounds you.

2. **Stay Physically Active:**

    Chopping wood, tending animals, or even just walking your property burns off stress and keeps your body — and mind — in good shape.

3. **Practice Gratitude:**

    Take a moment each day to notice what's going right. A good harvest, a sunny morning, a quiet moment by the fire — small wins matter more out here.

4. **Make Time for Creativity:**

    Read, write, carve, paint — anything that shifts your brain away from survival tasks and lets you create for the sake of creating.

5. **Stay Connected (When Possible):**

    Even off-grid, a radio check-in, a handwritten letter, or a visit from a neighbor goes a long way. Don't isolate yourself more than necessary.

6. **Learn to Be Okay with Solitude:**

    There's a difference between being alone and being lonely. Solitude can be a gift — a chance to think, breathe, and connect wit

Building a Support Network & Off-Grid Communities

Connection is one of the most underappreciated yet essential components of long-term off-grid survival. We are not destined to live in solitude forever, regardless of how independent we may be. You'll need assistance at some point, whether it's a helping hand during a difficult time, someone to trade with, or just the knowledge that someone else shares your lifestyle.

Developing a support system is about enhancing your independence, not sacrificing it. In addition to making difficult times easier, a strong off-grid community allows people to share resources, exchange talents, and build something larger and more resilient than any one individual.

Start Close to Home

Around you is where your first network begins. Even if your neighbors live far away, get to know them. You don't need to be best friends, but it helps to know who is close by, their abilities, and how they feel about mutual aid or bartering when things become challenging.

Before you need anything, introduce yourself. Building trust begins with a simple visit, lunch, or volunteering to assist with a project. In off-grid communities, trust is crucial.

Barter and Trade — Reviving Old Ways

Bartering is one of the most practical old-fashioned systems that come with living off the grid. Whether it's fresh veggies for mechanic work, eggs for firewood, or herbal salves for extra fencing supplies, establishing connections based on commerce forges bonds that money cannot purchase.

Begin modestly by asking what others need and sharing what you have in abundance. Without the need for a store, you'll eventually build a network of support where everyone wins.

Skill-Sharing and Knowledge Pools

Diversity of knowledge is one of a community's greatest assets. Your neighbor may be skilled at fixing solar panels, but you may be an excellent gardener. Another may have been a hunter, a herbalist, or a livestock farmer.

Trade abilities. Learn from one another. Organize unofficial workshops or get-togethers where people can exchange their knowledge. It makes everyone more prepared and fosters a sense of community.

Stay Connected, Even Off the Grid

You can still stay connected even if you don't have cell coverage or everyday internet access. Communication channels are maintained by ham radios, CB radios, shortwave radios, and even planned meetings at neighborhood gathering places.

Your network will be genuine, dependable, and prepared when you need it if you can check in, particularly during emergencies, hurricanes, or large projects.

DIY Project 14:
Make an Emergency Barter Kit

An emergency barter pack is one of the best methods to prepare for long-term off-grid survival. Cash may quickly lose value during a crisis, such as a natural disaster, supply chain breakdown, or economic collapse. But products and abilities? They will always be worth something.

When circumstances go hard, you can use a barter kit to trade for everything you need, such as food, gasoline, tools, or medical assistance. By building it today, you'll have trade-ready goods before you run out of options.

Tools & Materials

- A sturdy box, waterproof bag, or lockable container (easy to transport)
- Ziplock bags or small containers for organization
- Notepad and pen (to track trades if needed)
- **Optional:** fireproof bag or ammo can for added protection

Step 1 — Choose Barter Items

Focus on small, high-demand items that are easy to store, carry, and trade. Think about what you'd want if you ran out — and what others might need most.

Essential Barter Items to Include:

- **Lighters, waterproof matches, or ferro rods**
- **Batteries (AA/AAA)**
- **Small bottles of alcohol** (for disinfecting or trading)
- **Salt, sugar, coffee, tea**
- **Sealed medical supplies** (bandages, antiseptic wipes, over-the-counter pain relievers)
- **Fishing hooks and line**
- **Sewing needles and thread**
- **Soap bars or travel-size hygiene products**
- **A few packs of seeds** (heirloom, non-GMO if possible)
- **Ammo** (calibers common to your area — only if you're comfortable)
- **Candles or small solar lights**

Choose items that store well and don't spoil — you want this kit ready even years later.

Step 2 — Organize and Pack

Put related supplies, such as fire-starting, medical, and hygiene, in labeled bags or containers in your barter kit. Make sure everything is dry, safe, and convenient to reach quickly.

Ensure the container is sturdy and transportable; a heavy-duty plastic pail, waterproof backpack, or compact, lockable toolbox are excellent options.

Step 3 — Add a Simple Trade Ledger (Optional)

Tuck a notepad and pen inside. If you're bartering regularly, it helps to track what you've traded, what you're running low on, and any ongoing exchanges you're building with neighbors.

Step 4 — Store Securely and Review Annually

Keep your barter kit somewhere dry and secure but accessible if you need to leave quickly. Once a year, open it up, check expiration dates, swap out any degraded items, and restock what you've used or traded.

End-of-Chapter Checklist: Securing Your Long-Term Resilience

- ☑ **Doors and Windows Fortified** — Have you reinforced your main entry points with solid doors, strong locks, and window barriers to protect your home from intruders and wild animals?

- ☑ **Perimeter Fencing and Concealment in Place** — Is your property clearly defined with fencing, natural barriers, or strategic landscaping that adds privacy and protection?

- ☑ **Early Warning Systems Installed** — Have you set up motion sensors, alarms, or other systems to alert you of movement near your homestead — giving you time to react?

- ☑ **Guard Animals or Livestock Integrated** — Have you considered adding guard dogs or other animals that can help alert you to danger and provide another layer of defense?

- ☑ **Mental Resilience Habits Practiced** — Are you actively working on coping skills, daily routines, or creative outlets that help you manage isolation, stress, and the psychological demands of off-grid living?

- ☑ **Support Network or Community Contacts Established** — Have you reached out to nearby neighbors, like-minded homesteaders, or barter partners to strengthen your local support system?

- ☑ **Barter Kit Assembled and Ready** — Do you have a small stockpile of trade goods — essentials like lighters, medicine, tools, or food staples — ready to barter if cash loses value?

- ☑ **Skill Sharing and Knowledge Pool Growing** — Are you exchanging skills or knowledge with your community, building long-term relationships that help everyone thrive?

FINAL SECTION:
OFF-GRID READINESS & DIY PROJECTS

Building a lifestyle based on independence, understanding, and adaptability is more important than having every device or system in place when you're ready to live off the grid. The necessities—food, water, energy, health, safety, and mental toughness—have already been discussed. While each component is significant, they form a solid base that can withstand the benefits and hardships of independent living.

Putting those talents together and acting on them is the focus of this last portion. DIY projects are essential to living off the grid and aren't only a way to save money or keep occupied. Every system you create, tool and enhancement moves you closer to true preparedness. One project at a time, confidence is developed there as theory and practice collide.

From easy builds to long-term value-adding systems, this list of practical do-it-yourself projects can help you hone your abilities and improve your farm. These projects help ensure you're not just surviving off the grid but thriving, regardless of how you're starting out or optimizing your setup.

Chapter 15: Final Off-Grid Readiness Checklist & Action Plan

30-Day Off-Grid Transition Plan

Your Roadmap to Leaving the Grid with Confidence

Making the leap into off-grid living doesn't happen overnight — but with the right plan, you can break it down into manageable steps that build momentum. This 30-day transition plan will help you tackle the most critical areas — food, water, power, shelter, and mindset — while testing your systems before unplugging.

Each day is a small win, but by the end of the month, you'll have a solid foundation and a real taste of what off-grid life feels like.

Week 1 — Assess, Plan, and Prepare

Lay the groundwork and know exactly where you stand.

- ☑ **Day 1:** Take inventory of your current resources — water, food, fuel, tools, skills. Write it down.
- ☑ **Day 2:** Define your off-grid goals — temporary, seasonal, or permanent? Survival focus or full-time homestead.
- ☑ **Day 3:** Map your land or future site — water sources, sunlight patterns, and potential garden space.
- ☑ **Day 4:** Research your local regulations on water collection, waste management, and alternative energy.
- ☑ **Day 5:** List your critical systems — food, water, power, hygiene, heat. Rank what needs attention first.
- ☑ **Day 6:** Create a budget and prioritize spending — essential gear, tools, or projects come first.
- ☑ **Day 7:** Set up a basic off-grid notebook or binder — this will become your personal survival guide.

Week 2 — Water, Power, and Shelter Basics

Begin testing your independence from utilities.

- ☑ **Day 8:** Practice living one full day on stored water only. Track usage. Adjust your water plan.
- ☑ **Day 9:** Inspect or plan your rainwater collection setup. Start a DIY barrel system if possible.
- ☑ **Day 10:** Test an alternative cooking method — rocket stove, fire pit, or solar oven.
- ☑ **Day 11:** Calculate your daily energy use — what's essential? Test with a watt meter if possible.
- ☑ **Day 12:** Set up a small solar kit or portable battery pack. Practice running lights or small appliances.
- ☑ **Day 13:** Check your shelter — fix leaks, seal drafts, think about insulation or weather-proofing
- ☑ **Day 14:** Test your backup heat source — wood stove, propane, or solar heating. What needs improvement?

Week 3 — Food, Health, and Hygiene Systems

Work on securing your food supply and personal care systems.

- ☑ **Day 15:** Inventory your current food stores — aim for 30 days' worth of essentials. Start a preservation plan.
- ☑ **Day 16:** Practice a full day eating only from your pantry or garden. Note gaps.
- ☑ **Day 17:** Build or upgrade your composting toilet or waste management system.
- ☑ **Day 18:** Start a small garden or sprouting station — greens you can grow anywhere.
- ☑ **Day 19:** Assemble your first aid kit and herbal remedies — make one salve or tincture.
- ☑ **Day 20:** Do a "waterless hygiene" test — dry shampoo, natural toothpaste, sponge bath.
- ☑ **Day 21:** Practice basic hunting, trapping, or fishing skills — even if it's just reading or dry runs.

Week 4 — Security, Resilience, and Community

Test your readiness and strengthen the systems that keep you safe and sane.

- ☑ **Day 22:** Check your property for weak spots in fencing, doors, and windows.
- ☑ **Day 23:** Install a basic alarm or motion sensor system — even a DIY version.
- ☑ **Day 24:** Make your emergency barter kit — pack it with trade-ready supplies.
- ☑ **Day 25:** Spend 24 hours without grid power — no phones, internet, or AC. What breaks first?
- ☑ **Day 26:** Practice a self-reliance skill — fire starting, water purification, herbal medicine, or foraging.
- ☑ **Day 27:** Build your network with one like-minded neighbor or online community.
- ☑ **Day 28:** Review your mental resilience plan — hobbies, routines, or coping strategies for tough days.

Days 29-30 — Test and Adjust

Final push — simulate an entire off-grid weekend and fine-tune what's missing.

- ☑ **Day 29:** Spend two full days as off-grid as possible — no grid water, power, or outside help
- ☑ **Day 30:** Review what worked and what didn't. Update your off-grid plan. Set your next 30-day goal — expand systems, add animals, or upgrade the shelter.

6-Month Homesteading Development Plan

Building a Strong, Self-Sufficient, Off-Grid Life — One Month at a Time

You don't become an expert homesteader overnight. It's a layer-by-layer lifestyle that involves learning from your land as the seasons change, developing new skills, and testing systems. This 6-month strategy allows you to develop each crucial area without feeling overburdened by breaking the process into small monthly objectives.

By the end, you will have more than simply a piece of property; you will have the foundation of a robust, fruitful homestead that will sustain your long-term off-grid lifestyle.

Month 1 — Land, Layout, and Resource Planning

Start by knowing your space and what you're working with.

- ☑ Walk your land daily — observe sunlight patterns, water flow, and wildlife activity.
- ☑ Map out your homestead — garden plots, livestock areas, water access, and shelter sites.
- ☑ Inventory natural resources — firewood, wild plants, building materials
- ☑ Test your soil — check quality and plan amendments
- ☑ Begin clearing small garden space or work area
- ☑ Set short-term and long-term goals: what do you want functional in 6 months?

Month 2 — Water Systems and Storage

Secure your most important resource — reliable water access.

- ☑ Set up or upgrade rainwater collection (gutters, barrels, filtration)
- ☑ Dig or inspect wells if applicable; test water quality
- ☑ Build a basic greywater system for garden use
- ☑ Learn and practice multiple water purification methods
- ☑ Create a water usage plan — daily needs for people, animals, and plants
- ☑ Stock up on water containers or build an emergency storage system

Month 3 — Food Production: Gardens, Animals, and Preservation

Start growing, raising, and preserving your food supply.

- ☑ Expand or plant your first serious garden bed (raised beds, in-ground, or containers)
- ☑ Begin composting kitchen scraps and animal manure
- ☑ Choose one small livestock species to start — chickens, rabbits, or ducks
- ☑ Build housing for animals: coop, hutch, or fencing
- ☑ Practice food preservation: canning, dehydrating, fermenting
- ☑ Create a seed bank with heirloom, non-GMO seeds
- ☑ Start a foraging journal — identify local edible plants

Month 4 — Energy, Heating, and Fuel Systems

Start reducing reliance on outside power sources.

- ☑ Install or expand your solar system or alternative energy source
- ☑ Practice using wood stoves, rocket stoves, or solar ovens for cooking
- ☑ Split and stack firewood or secure your winter fuel source
- ☑ Build or improve an outdoor cooking area
- ☑ Set up backup power — battery banks, generators, or solar chargers
- ☑ Review energy needs — plan what can be cut or shifted off-grid

Month 5 — Home Skills, Repairs, and Self-Reliance Projects

Build practical skills that keep your homestead running smoothly.

- ☑ Learn basic carpentry — build or repair fences, animal shelters, or storage.
- ☑ Make natural cleaners, soap, or toothpaste from scratch
- ☑ Practice herbal medicine — tinctures, salves, teas
- ☑ Repair clothes, mend gear, and practice sewing skills
- ☑ Test your first aid skills — wound care, splinting, treating burns
- ☑ Expand your food storage — root cellar, pantry shelves, or cold storage

Month 6 — Security, Community, and Long-Term Resilience

Strengthen your systems and relationships for long-term survival.

- ☑ Walk your property lines — reinforce fences, gates, and natural barriers
- ☑ Install basic security measures — motion sensors, cameras, alarms
- ☑ Build or stock an emergency barter kit
- ☑ Make contact with local homesteaders, farmers, or barter groups
- ☑ Practice a full 48-hour off-grid test — no grid power, outside help, or store trips
- ☑ Create or review your homestead emergency plan — fire, weather, medical
- ☑ Assess your progress and set your following 6-month goals

Final Off-Grid Readiness Scorecard — Self-Assessment Quiz

How Ready Are You to Live Off the Grid?

It's crucial to stop and assess your level of preparedness at any point in your off-grid adventure. Being honest with yourself, recognizing your strengths, and identifying areas for improvement are the goals of this scorecard, not perfection.

After reading each question, provide yourself with:

- ☑ **3 points —** Confident and Ready
- ☑ **2 points —** Partially Ready / In Progress
- ☑ **1 point —** Needs Work

Grab a pen or notebook, total your score, and see where you stand.

Water Security

1. Do you have at least two reliable water sources (rain collection, well, stream, stored supply)?
2. Can you purify water without electricity — using filters, boiling, or DIY methods?
3. Is your water system protected from seasonal changes or contamination?

Food Security & Preservation

1. Do you produce or gather a significant portion of your food (gardening, hunting, livestock)?
2. Do you have a system to preserve food long-term — canning, drying, fermenting?
3. Can you cook without grid power — using a wood stove, solar oven, or open fire?

Energy & Heating

1. Is your power supply independent — solar, wind, or other renewable systems?
2. Do you have a backup heat source— wood stove, propane heater, or thermal mass?
3. Can you run basic tools or equipment off-grid if needed?

Waste Management & Hygiene

1. Do you have a composting toilet or alternative waste system ready to use?
2. Can you manage hygiene and sanitation with limited or no running water?
3. Have you made or stocked natural hygiene products (soap, toothpaste, cleaners)?

Medical & First Aid Preparedness

1. Do you have a fully stocked first aid kit ready for off-grid emergencies?
2. Have you learned basic first aid or herbal remedies for common injuries or illnesses?
3. Do you have backup medical plans for serious health emergencies?

Security & Home Defense

1. Is your home and property secured with reinforced doors, windows, or fencing?
2. Do you have an early warning system — motion sensors, alarms, or animals?
3. Have you practiced emergency drills or know what to do if threatened?

Mental Resilience & Community

1. Do you have a daily routine or hobbies to cope with isolation or stress?
2. Have you built connections with neighbors, a barter network, or an off-grid community?
3. Do you have a barter kit or supplies set aside for trade?

Self-Sufficiency & Skills

1. Can you build or repair basic structures — fencing, animal shelters, and simple plumbing?
2. Do you know how to forage, hunt, trap, or fish for food if necessary?
3. Have you tested yourself with off-grid challenges — 24-hour or 48-hour survival runs?

Score Yourself

→ **60–72 Points: Off-Grid Ready —**
You've built a solid foundation. Keep fine-tuning, but you're ready to thrive off-grid.

→ **45–59 Points: On Your Way —**
You're progressing well. Focus next on any weak areas you spotted.

→ **30–44 Points: Early Stage —**
You've got a good start, but there's work to do. Tackle one system at a time.

→ **Below 30 Points: Vulnerable —**
Time to roll up your sleeves. Start with water, food, and energy — your survival priorities.

Final Refinements: Strengthening Your Off-Grid Plan

You've established a strong basis for off-grid living, from testing your systems and abilities to obtaining your basic necessities. Off-grid living isn't static, though. You're always developing, modifying, and refining as you develop, learn, and overcome new difficulties.

In this concluding piece, we'll go deeper into the experiences of those who have traveled this journey, providing case studies and lessons discovered the hard way. Additionally, you will learn about some of the newest off-grid technology that can simplify, secure, and improve your quality of life, such as advanced water purification systems and contemporary battery storage options.

We'll go over more advanced security techniques and community-building advice to help you establish an independent homestead and a support system to increase your long-term resilience.

You'll conclude this adventure with a self-assessment scorecard, which will allow you to evaluate your preparedness, acknowledge your progress, and determine your future areas of concentration.

Living off the grid is a way of life, not a place to visit. The purpose of these last adjustments is to assist you in creating a homestead that is both sustainable and prepared to prosper in the real world, regardless of what the future brings.

More Real-World Examples & Case Studies

Learning from Those Who've Lived It

The beauty of off-grid travel is that no two trips look exactly alike. Every family, homestead, and plot of land has its own set of difficulties, revelations, and lessons. When theory meets practice, off-grid living may be very different and enjoyable, as demonstrated by the real-world examples below.

Case Study #1 — A Family's Journey from Suburbia to 10 Acres Off-Grid

Location: Northern Arizona, USA

Focus: Water scarcity, homeschooling, solar power

After years of suburban living, the Johnson family of five sold everything and bought 10 remote acres. Their biggest challenge was water — no well, no creek, just raw land. They relied on rainwater collection and hauling water while saving for a well. Solar panels power essentials like lighting, a fridge, and homeschooling devices.

Key takeaway: *"We thought the hardest part would be power. It was actually water. Hauling it in gave us an entirely new appreciation for every drop."*

Case Study #2 — Single Woman's Mountain Cabin & Mental Resilience

Location: British Columbia, Canada

Focus: Living solo, handling isolation, heating with wood

Emma left city life in her early 40s to build a 600-square-foot cabin deep in the mountains. Her biggest struggle wasn't the cold winters or physical labor — it was managing isolation and loneliness during long winters. She built a support network of local farmers and loggers, traded firewood for homegrown vegetables, and learned to create routines that kept her mentally strong.

Key takeaway: *"The loneliness hits hard if you're not ready. I found my sanity in simple things — feeding the chickens, morning coffee, writing a journal."*

Case Study #3 — Desert Homestead Powered by Hybrid Energy

Location: New Mexico, USA

Focus: Hybrid solar-wind power system, desert gardening, water catchment

The Ramirez couple chose the desert for privacy and sunlight but quickly realized solar alone wasn't enough. They added a small wind turbine, giving them power during stormy nights when solar faltered. Gardening required shade structures and drip irrigation from rain catchment tanks.

Key takeaway: *"Desert life taught us flexibility. It's not about sticking to one system — it's about layering solutions until things work."*

Case Study #4 — Multi-Family Off-Grid Community Experiment

Location: Rural Vermont, USA

Focus: Shared resources, community resilience, barter systems

Three families bought adjoining land and built separate small homes but shared a large garden, livestock, and a community barn. They ran joint solar power and water systems, traded skills, and created a barter economy for labor, food, and goods — allowing each family to specialize.

Key takeaway: *"None of us could have done this alone. Together, it became not just survival — it became a community."*

Case Study #5 — Aging Couple's Adaptation to Off-Grid Life

Location: Tasmania, Australia

Focus: Downsizing, adapting for age, prioritizing low-maintenance systems

In their 60s, the Martins built a simple off-grid cabin focused on ease of use as they aged — raised garden beds, gravity-fed water, solar with battery backup, and low-maintenance animals like chickens. Their approach centered around "less is more," focusing on what truly mattered.

Key takeaway: *"Off-grid doesn't have to mean grinding labor daily. We designed it to slow down and remain sustainable as we age."*

More Modern Off-Grid Tech & Alternative Energy Storage

Smart Tools for a More Efficient, Sustainable Homestead

While off-grid living often draws from traditional skills, modern technology has come a long way in making this lifestyle more sustainable, efficient, and comfortable. Today, there are smarter, more reliable ways to generate, store, and manage your energy — all designed with off-grid independence in mind.

Here's a look at some of the latest tools and systems making a real difference on modern homesteads:

1. Lithium Iron Phosphate (LiFePO4) Battery Banks

Forget old-school lead-acid batteries — many off-gridders are investing in LiFePO4 battery systems today. These batteries last longer (up to 10 years), handle deep discharges better, and are safer indoors.

- **Benefits:** Lighter, longer-lasting, more stable, and zero maintenance
- **Best use:** Solar or wind energy storage, powering your home, tools, and emergency systems

Popular brands like **Battle Born, Renogy,** or **Bluetti** offer modular battery banks that scale with your system.

2. Portable Power Stations (Solar Generators)

Off-gridders' backup power management is evolving because of the introduction of small, rechargeable solar generators, such as the Jackery, EcoFlow Delta, or Bluetti models. These portable boxes integrate solar inputs, inverters, and lithium batteries.

- **Benefits:** Plug-and-play design, portable, perfect for emergency power, tools, or RVs
- **Best use:** Running essential appliances during outages, powering tools in remote areas, charging devices

3. Smart Inverters with App Monitoring

Modern **hybrid inverters** now come with Bluetooth or Wi-Fi capability, letting you monitor energy production, battery levels, and usage from your phone.

- **Brands to watch: Victron Energy, Growatt, Schneider Electric**
- **Why it matters:** Real-time data helps you avoid surprises and manage your power smarter — especially during tough weather or winter.

4. Flexible & Rollable Solar Panels

Not all solar has to be rigid and roof-mounted. **Flexible or rollable solar panels** now give you more freedom to power tools, charge devices, or even set up a mobile solar array wherever you need.

- **Best use:** Campsites, RVs, backup power, powering electric fences or small pumps
- **Example:** Renogy's 100W flexible panels

5. Water-Powered Backup Generators (Micro Hydro)

For homesteads lucky enough to have flowing water, **micro-hydro generators** are one of the most reliable 24/7 off-grid power sources — often outperforming solar during cloudy seasons.

- **Benefits:** Constant power, even at night or during storms
- **Challenge:** Requires consistent water flow and setup investment

6. Solar-Powered Refrigeration & Freezers

Refrigeration used to be a huge energy drain — not anymore. Off-griders now have access to **DC-powered chest freezers and fridges** explicitly designed for solar setups.

- **Brands to watch: Sundanzer, EcoSolarCool, Dometic**
- **Why it matters:** Efficient cold storage protects your food security without draining your battery bank

7. Biochar & Gasifier Systems for Fuel and Soil Health

Forward-thinking homesteaders are exploring **biochar kilns** and **wood gasifiers** — systems that turn wood waste into fuel or soil-building charcoal.

- **Benefits:** Turns waste into usable energy or carbon-rich soil amendments
- **Best use:** Heating, cooking, improving garden yields

8. Rainwater Harvesting & Smart Water Filters

New filtering systems — **Berkey, Sawyer, and Lifestraw Community Filters** — offer high-capacity filtration for off-grid homes. Paired with large **IBC totes or smart rain collection barrels,** they create reliable water systems with minimal maintenance.

Expanded Security & Community-Building Strategies

Strengthening Your Defense and Support Systems for Long-Term Survival

Living off the grid requires you to be entirely accountable for your health and safety. It also entails understanding that genuine resilience is strengthened by developing reliable relationships around oneself. Since both are necessary for long-term survival and success, we'll detail doable strategies to increase your physical security and community networks in this part.

Advanced Security Strategies for Your Homestead

Securing your off-grid home goes beyond locks and fences. Real security is about layers — slowing down intruders, buying yourself time, and deterring threats before they reach your doorstep.

1. Layered Defense Approach

Think like a predator — if it's easy, they'll try. If it looks risky, they'll move on. Layer your defenses:

- **Outer Layer:** Natural barriers (thick hedges, thorn bushes, berms), fencing, hidden trails to avoid creating obvious paths
- **Middle Layer:** Motion sensors, cameras, tripwire alarms, or solar-powered floodlights
- **Inner Layer:** Reinforced doors, window shutters, safe rooms, and emergency communication setups

2. Dummy Systems and Decoys

Sometimes, what *looks* like a defense is as effective as the real thing:

- Install a "dummy" camera system or signs warning of surveillance
- Fake trail cameras in the woods can make people think twice about sneaking up
- Empty animal traps along property lines act as psychological deterrents

3. Guard Animals as Early Warning

Don't underestimate the value of animals in your security plan:

- **Geese** are noisy and aggressive — natural alarms
- **Livestock guardian dogs (LGDs)** protect animals and alert you to movement
- Even donkeys and llamas can defend livestock from predators

4. Concealed Observation Points

Have one or two spots where you can quietly watch your property — a tree stand, a small lookout, or a tucked-away hilltop. It's not about being paranoid — it's about staying aware when it matters.

5. Practice Drills and Scenario Planning

Run mock scenarios:

- What if someone walks onto your land?
- What if a storm takes out your communications?
- How fast can you secure your home or call for help?

Practicing keeps your reflexes sharp and shows you where your weak points are.

Building Community for Long-Term Resilience

Having trustworthy people makes off-grid living much easier and less isolating. A big gathering isn't always a sign of community. Sometimes, you only need a few trustworthy people you can rely on in difficult times.

1. Start Small — Build Local Relationships

Learn about local farmers, homesteaders, and even seasonal hunters. Offer assistance, exchange resources, and share expertise; initially, there are no conditions. Although it takes effort, trust eventually pays dividends.

2. Skill-Sharing Gatherings

Host or attend small gatherings focused on:

- Seed swapping
- Foraging walks
- First aid workshops
- Butchering days or harvest parties

It's a low-pressure way to build bonds while learning from each other.

3. Build a Barter Network

Before a crisis arises, begin trading products and services. Exchange work for fuel, eggs for firewood, or herbs for repairs. As time passes, this decreases your need for money and fosters trust.

4. Communication Plans with Neighbors

Create simple check-in systems:

- Radio check once a week
- Shared emergency signals (flag system, rifle shot count, pre-arranged code words)
- Plan for sharing news or warnings if things go sideways

5. Form Mutual Aid Pacts

A "we've got each other's back" agreement with a few neighboring homesteads doesn't have to be official. Still, it can save lives during medical crises, fires, natural catastrophes, or security threats.

Final Off-Grid Readiness Scorecard — Self-Assessment Quiz

Check Your Progress — How Ready Are You to Thrive Off the Grid?

This self-assessment will help you honestly measure where you stand on your off-grid journey. There's no perfect score — this is about identifying your strengths and spotlighting areas where you might need to grow or refine your systems.ection.

For each statement below, score yourself:

- ☑ **3 points** — Fully Ready
- ☑ **2 points** — In Progress
- ☑ **1 point** — Needs Work

Tally your points at the end to see where you stand.

Water Security

1. I have two or more reliable water sources (well, rain collection, stream, storage).
2. Without grid power, I can purify water with filters, boiling, or DIY systems.
3. I've tested my water systems for seasonal changes (droughts, freezes).

Food Production & Storage

1. I grow, raise, hunt, or forage a significant portion of my food.
2. I have food preservation systems (canning, dehydrating, smoking).
3. I've stored enough food for at least 3-6 months of self-reliance.

Energy & Power

1. My home has an independent power system (solar, wind, hydro, or hybrid).
2. I have backup energy storage (batteries, fuel, or alternative systems).
3. I can cook, heat, and power essentials off-grid year-round.

Shelter, Waste, and Hygiene

1. My home or shelter is reinforced, weatherproof, and secure.
2. I have a functioning composting toilet or alternative waste system.
3. I can manage hygiene, laundry, and sanitation off-grid with limited water.

Medical Preparedness

1. My first aid kit is fully stocked and includes herbal or natural remedies.
2. I'm confident handling medical emergencies like cuts, burns, or infections.
3. I have a plan for handling serious injuries or medical needs without immediate outside help.

Security & Home Defense

1. My property is secured with fencing, barriers, or surveillance.
2. I have early warning systems (dogs, motion sensors, natural defenses).
3. I've practiced drills or scenarios for home defense and emergency situations.

Mental Resilience & Community

1. I have daily routines or mental practices that help me cope with isolation or stress.
2. I'm connected to a local community or network I can count on.
3. I have a barter kit and trusted partners for trade or mutual aid.

Practical Skills & Adaptability

1. I can forage, hunt, fish, or trap to supplement my food supply.
2. I can build or repair basic structures (shelters, fences, animal enclosures).
3. I've completed at least one off-grid test run (24-48 hours fully self-reliant).

Your Total Score: _____ / 72

How to Read Your Score:

→ **60–72 Points — Off-Grid Ready:**

You're well-prepared, with solid systems in place. Keep refining, but you're ready to thrive.

→ **45–59 Points — Strong Progress:**

You're on your way. Focus on shoring up your weakest areas — resilience is about covering every base.

→ **30–44 Points — Early Stages:**

You've started the journey but need more work in core areas. Tackle water, food, and energy first.

→ **Below 30 Points — Vulnerable:**

Your foundation needs strengthening. Start with essential survival systems, then build outward.

The Last Word about Off-Grid Living

When you take a step back and consider all we've discussed in this guide, you'll see that off-grid life is more than just a set of survival skills or projects. It's a way of thinking. A return to the notion that people may create fulfilling lives from the bottom up if they are freed from convenience and reliance.

The goal of this book has been to illustrate the way by comprehending the significance of each system, not merely by providing instructions or diagrams. Security of water. Food self-sufficiency. Sustainable energy. Management of waste. Safety. Mental toughness. These aren't distinct subjects. They assemble parts of a life meant to cooperate with nature rather than resist it constantly.

As has always been the case, water was the first important priority. There would be no comfort, much less survival, without it. We dissected every facet, including greywater recycling, well excavation, purification techniques, and rainwater collection systems. This is because your off-grid homestead's water system is its lifeblood. If you can't keep water flowing to your crops, animals, and yourself, then no amount of glistening solar panels or a beautiful garden is worth anything. The water required to survive and thrive shocked many first-timers and will probably surprise you. The difference between battling through a dry summer and navigating it with assurance is preparation.

We then addressed energy. Modern off-grid living has made possible possibilities our ancestors could only imagine, thanks to solar panels, wind turbines, micro-hydro, and battery storage. These days, power can be obtained without a power company. Light without the grid. Even though the closest store is fifty miles away, food is still in the freezer. However, the lesson is clear: adding more devices isn't the answer to energy independence. It all comes down to identifying your needs, reducing waste, and creating systems that effectively address them. Any day, a compact, well-thought-out solar array with energy-efficient appliances and LED lighting is preferable to a large array with inefficient electronics.

Perhaps more than any other topic in this book, food security shows how independence changes a person. Your perspective on the land is altered when you grow your own food. Rich soil is created from what used to be dirt. A patch of weeds can now be transformed into a possible salad bed, medicine cabinet, or habitat for pollinators. Every project adds another layer of food resiliency, from permaculture design and raised bed gardening to rearing goats, chickens, bunnies, and bees. And eating enough isn't the only thing that matters. Better food is the goal, whether it's honey collected from your own hives, eggs collected still warm from the coop, or meals taken directly from your soil. Few things remind you that you're alive more than feeding yourself with your hands.

Preservation and storage were next since having food today means little if you can't make it last. Smoking, canning, drying, fermenting—all age-old practices brought into the modern homestead because survival doesn't happen in seasons. It's year-round. The world doesn't care if your garden was fantastic this summer if you didn't put any of it away for winter.

Waste management and hygiene earned their position on these pages because no one talks about it enough, yet it's vital. A functioning composting toilet, greywater system, and strong hygiene strategy preserve your health, land, and water. There is no anonymous sewer system to carry your issues away. What you build, you live with. When done effectively, trash becomes a resource, nourishing gardens and rebuilding soil instead of contaminating streams.

Next came security, a topic that is inevitable with off-grid living. Protecting what you build is a component of being self-reliant, not because there are many villains in the world. Adding guard animals, installing perimeter alarms, and strengthening doors and windows should all be motivated by a sense of worthiness rather than fear. You move from isolation to resilience by incorporating mutual aid agreements, barter networks, and community-building techniques. A network of neighbors who individually contribute resources and abilities makes you stronger than ever.

However, a deeper theme permeates our discussion: the psychological and emotional aspects of living off the grid. This life is more than a set of tasks. It is lengthy periods of silent, difficult decisions, failures, and the gradual process of realizing how to live with less. More important than any equipment in your garage is mental toughness. Self-sufficiency can be romanticized until the first winter storm strikes, or you're left wondering what went wrong while looking at a flooded garden. The best farmers or constructors aren't usually the ones that survive. They adjust, find joy in the simplicity of a well-done task, and laugh when things go wrong.

It's not a checklist that I want you to take away from this book. It's the understanding that living off the grid isn't a binary choice. You don't become completely independent one morning. You work on it, season by season, project by project. Perhaps you begin with a garden. Perhaps you can construct a rainwater system or install solar panels. Each component supports the overall while simultaneously standing alone.

Since perfect self-sufficiency does not exist, there is no such thing as an ideal off-grid homestead. You'll always barter for something, trade for something, or learn something later. And that's all right. Redefining your relationship with the world is more important than trying to avoid it. Becoming active in your survival, comfort, and well-being rather than a passive consumer.

If I've discovered anything, those who are patient, inquisitive, and unyielding are rewarded in this life. It's not glitzy. When you fix your plumbing or harvest from a garden you constructed, there won't be any camera crew to cheer you on. You'll know, though. You'll sense it. That calm pride comes from knowing that you can support yourself and your loved ones in a society that appears to have forgotten what that means.

You won't freak out when the electricity lines fall, store shelves are empty, or the world becomes unpredictable, as it always does. You will adapt. You'll solve problems. You'll use the resources you manage, the systems you create, and the abilities you acquire.

That's the main focus of this book. Not only getting by. Not merely checking boxes. But prospering according to your terms. Establishing a life in which you fully and deeply own your food, water, energy, health, and security rather than outsourcing them.

You're already on the right track if you've read this far. Continue. Continue to construct. Continue to learn. And keep in mind that independence isn't about being alone. It's about building a life full of community, expertise, and skills so you know you're prepared for everything coming your way.

Off-grid living is that. Not a retreat. Not a way out. A return, however—to oneself, to the soil, to a way of life that has always been there for those who dare to reclaim it.

IMAGES USED UNDER LICENSE FROM SHUTTERSTOCK.COM

PAGE 1	BUBLIKHAUS	PAGE 61	PAUL D'ARVILLE
PAGE 5	INNA DODOR	PAGE 62	WARUJ KUNNAMAUNG
PAGE 6	WILLIAM EDGE	PAGE 63	TOPPERSPIX
PAGE 7	MOOMIN201	PAGE 64	SANDRA KOKA
PAGE 10	WILLIAM EDGE	PAGE 68	SHEBEKO
PAGE 19	DELOVELY PICS	PAGE 75	NATALLIASKN
PAGE 27	DAINIS ZVINGULIS	PAGE 77	MARC ELIAS
PAGE 33	SAJID1264	PAGE 82	PAT_HASTINGS
PAGE 40	PIYASET	PAGE 87	SKYLIZ
PAGE 47	IFISTUDIO	PAGE 89	KANDYBKA ALINA
PAGE 49	MS SELL PIC	PAGE 101	JAVIER MENDOZA OLMOS
PAGE 54	ADAMIKARL	PAGE 102	ALEKSEY MATRENIN
PAGE 60	OLEG MIKHAYLOV		

www.ingramcontent.com/pod-product-compliance
Lightning Source LLC
Chambersburg PA
CBHW051327110526
44582CB00003B/76